S0-BDL-355

Nevada Environmental Issues

Edited by

David A. Charlet

Community College of Southern Nevada

With Written Contributions by

Lesley Argo, David A. Charlet, Therese N. Charlet,
James S. Coleman, Ed Eschner, Lynn K. Fenstermaker,
David M. Hassenzahl, Travis E. Huxman, Robert S. Nowak,
J.D. Pellock, Marc L. Rigas, Amina Sadik, Denise Signorelli,
Stanley D. Smith, Randy Smith, and Shane Snyder

KENDALL/HUNT PUBLISHING COMPANY
4050 Westmark Drive Dubuque, Iowa 52002

Cover photos from top left to right.

Ash Meadows National Wildlife Refuge by David A. Charlet
David A. Chalet at Mt. Rose Wilderness Area by Dean Hagen
McCullough Station, Ed Dorado Valley by David A. Charlet
Phainopepla at Las Vegas Valley by David A. Charlet
Atom bomb explosion courtesy U.S. Department of Energy
Moth at Mt. Rose Wilderness Area by Dean Hagen
Summit Lake by David A. Charlet
Oenothera by David A. Charlet
Ash Meadows National Wildlife Refuge by David A. Charlet
FACE circle, Nevada Test Site by Lynn Fenstermaker

Copyright © 2002 by Kendall/Hunt Publishing Company

ISBN 0-7872-9488-8

All rights reserved. No part of this publication may be reproduced, stored in a retrieval system, or transmitted, in any form or by any means, electronic, mechanical, photocopying, recording, or otherwise, without the prior written permission of the copyright owner.

Printed in the United States of America
10 9 8 7 6 5 4 3 2 1

Contents

About the Authors

David A. Charlet

David Charlet received his M.S. in Biology and his Ph.D. in Ecology, Evolution, and Conservation Biology from the University of Nevada, Reno. He currently is a Professor at the Community College of Southern Nevada in the Las Vegas Valley, where he teaches ten biology and environmental science classes each year. David Charlet's research is acutely focused on the natural history of the Great Basin and Mojave Desert. Dr. Charlet has worked in most of Nevada's 314 named mountain ranges, and has mapped and written a reference book on the conifers of Nevada. Dr. Charlet mapped the vegetation of the Carson Range and much of the eastern Sierra Nevada under a National Performance Review grant sponsored by Vice President Gore's Committee on Reinventing Government. Currently, his activities include basic research for the Clark County Multi-Species Habitat Conservation Plan.

David A. Charlet, Ph.D.
Community College of Southern Nevada
Department of Biological Sciences H3C
700 College Drive
Henderson NV 89015
Email: david_charlet@ccsn.nevada.edu

Lesley Argo

Lesley Argo received her B.A. in Political Science in 1995 from the University of New Mexico. In 2001, she earned her M. S. in Environmental Science from the University of Nevada, Las Vegas, where she studied issues surrounding public lands in the rural counties of the state. Lesley is currently completing course work for her doctorate, and plans to continue her research and writing about the people and public lands of rural Nevada.

Lesley Argo, M.S.
Department of Environmental Studies
University of Nevada, Las Vegas
Box 454030
4505 S. Maryland Pkwy.
Las Vegas, NV 89154-4030
Email: lesleyargo@juno.com

Therese N. Charlet

Therese Charlet received her M.S. in Cellular and Molecular Biology from the University of Nevada, Reno in 1994. She is currently the Data, Grants, and Resource Manager for the Nevada Desert Research Facility (NDRC). Ms. Charlet is involved in many areas of the global climate change research underway at the center and is the author of the NDRC website.

James S. Coleman

James Coleman received his M.S. and his Ph.D. (1987) from Yale University. He is currently a Research Professor and Vice President at the Desert Research Institute in Reno, Nevada. Dr. Coleman has an active research program that investigates a wide spectrum of global change issues, including nutrient dynamics, carbon balance, and thermotolerance.

Ed Eschner

Ed Eschner is Professor of Geology in the Physical Sciences Department at the Community College of Southern Nevada.

Lynn K. Fenstermaker

Lynn Fenstermaker received her M.S. in Agronomy at Pennsylvania State University in 1986. She is currently a remote sensing scientist at the Desert Research Institute in Las Vegas and the Director of the Nevada Desert Research Center. Her research interests focus on using remote sensing techniques to quantify changes in vegetation under climate change scenarios. Ms. Fenstermaker will receive her Ph.D. from the University of Nevada, Las Vegas, in 2002.

Lynn Fenstermaker
Desert Research Institute
755 E. Flamingo Rd.
Las Vegas, NV 89119
Email: lynn@dri.edu

David M. Hassenzahl

David Hassenzahl is an Assistant Professor of Environmental Studies at the University of Nevada, Las Vegas. After earning his undergraduate degree in Paleontology and Environmental Science, he did environmental compliance work for five years in both the public and private sectors in the San Francisco area. Dr. Hassenzahl earned his Ph.D. in Public Affairs in June of 2000 from Princeton University, researching the role of risk analysis in public policy making. His current research includes Nevada's nuclear legacy and future, applications of risk analysis for decision-making, and risk analysis education. He is co-author of *Should We Risk It?*, a textbook on risk analysis for decision-making available in English and Japanese, and creator of the Risk Analysis Teaching and Learning (RATL) web site.

David M. Hassenzahl, Ph.D.
Department of Environmental Studies
University of Nevada, Las Vegas
Box 454030
4505 S. Maryland Pkwy.
Las Vegas, NV 89154-4030
Email: david.hassenzahl@ccmail.nevada.edu

Travis E. Huxman

Travis E. Huxman received his M.S. in Biology at California State University, San Bernadino, and his Ph.D. in Biology at the University of Nevada, Las Vegas in 1999. He is currently an Assistant Professor in Ecology and Evolutionary Biology at the University of Arizona in Tucson. Dr. Huxman's research interests are broadly focused on reproductive performance characteristics of desert plants. He also is actively involved in research at the Nevada Desert Research center.

Robert S. Nowak

Robert Nowak received his M.S. in Range Ecology (1980) and his Ph.D. in Range Ecology (1984) from Utah State University. He is currently a Professor in Environmental and Resource Science at the University of Nevada, Reno, where he teaches a variety of environmental and ecology courses. He also serves on the Editorial Board of the *Western North American Naturalist* journal and is a panel member and *ad hoc* reviewer for federal agencies (DOE, NSF, USDA, and USGS). His active research program investigates belowground plant processes and water balance in desert ecosystems.

Robert Nowak
Department of Environmental and Resource Science
University of Nevada, Reno
Reno NV, 89557
Email: nowak@cabnr.unr.edu

J.D. Pellock

J.D. Pellock received his B.S. from California State University, Stanislaus, and his M.S. in Chemistry at California State University, Fresno. J.D. is Professor of Chemistry in the Physical Sciences Department at the Community College of Southern Nevada. Previous work includes environmental chemistry, emissions research, industrial waste inspection, and environmental consulting.

Email: jd_pellock@ccsn.nevada.edu

Marc L. Rigas

Marc Rigas received a Ph.D. in Bioengineering from the Pennsylvania State University in 1997, where he studied the effects of air pollutants on respiratory function in humans. Since 1997, Dr. Rigas has been with the U.S. Environmental Protection Agency, at the National Exposure Research Laboratory in Las Vegas. His research interests continue to emphasize the impacts of environmental contaminants on human physiology and health. He serves on a federal coordinating committee of a national clinical study of environmental effects on children's health and development. Dr. Rigas has broad scientific and societal interests, teaching part-time in both biology and environmental science at the Community College of Southern Nevada and the University of Nevada-Las Vegas.

Marc L. Rigas, Ph.D.
U.S. EPA (HERB)
P.O. Box 93478
Las Vegas, NV 89193-3478
Email: rigas.marc@epa.gov

Amina Sadik

Dr. Amina Sadik is originally from Casablanca, Morocco. She completed her undergraduate degrees at the University of Mohammed V in Rabat, Morocco, in 1982. She earned her M.S. and D.E.A. in Plant Physiology and Food Processing in 1984. Her Ph.D. focus was on Phytopathology. Back in

Morocco as a tenured assistant professor she taught for five years, Phytopathology, Biology, and Physiology. Dr. Sadik came to the USA in August 1994. She worked for two years on a DNA library for adventitious root genes in the Biology Department at the University of Nevada, Las Vegas. In 1996, she worked on breast cancer research with Stephen W. Carper, Ph.D., Director of the Cancer Institute. Dr. Sadik is planning to use her knowledge of both animals and plants to find an adjuvant therapy using plant extracts in cancer treatment. Meanwhile, as soon as she felt confident enough in her English, Dr. Sadik went back to her first love, which is sharing her knowledge with others: teaching! She is currently an instructor at UNLV, where she teaches Clinical Chemistry, Biochemistry, and Organic Chemistry. She also supervises Clinical Practicum of Chemistry and Immunology. At the Community College of Southern Nevada, she teaches Environmental Science (online and in the classroom) and Biology. Finally, at the University of the Pacific, she facilitates Nutrition, Biology, Environmental Issues and Ethics, and Environmental Science. In her spare time, she enjoys cooking ethnic foods, swimming, walking, skiing, reading, and traveling as often as she can. English is her fourth language, and she is working on her fifth, Spanish. Since her arrival in Las Vegas, she has been learning the language and finding her niche in the area of research, while also being a mother, a wife, and making her house a home. She loves the Las Vegas climate, the beauty of the desert, and the proximity to the mountains.

Denise Signorelli

Denise Signorelli, Ph.D., is a full-time Biology and Environmental Science instructor in the Biological Sciences Department at the Community College of Southern Nevada.

Randy Smith

Randy Smith earned his B.S. from Humboldt State University in Fisheries. He earned a M.S in Biology from University of Nevada, Reno, and his Doctor of Arts degree in Biology from Idaho State University. Dr. Smith taught Environmental Science and Biology classes at the Community College of Southern Nevada for nine years. He has also taught classes at University of Nevada, Reno, Idaho State University, and California State University-Bakersfield. He is currently an adjunct instructor at the College of Southern Idaho. Dr. Smith's interests are in development of non-biology major courses and curriculum.

Stanley D. Smith

Stan Smith received his M.S. in Biology at New Mexico State University and his Ph.D. in Botany at Arizona State University (1981). His current position is Professor of Biology at the University of Nevada, Las Vegas where he has an active research program in global climate change biology. He teaches a variety of ecology courses, serves on the Editorial Board of the journals *Ecology* and *Ecological Monographs*, and is a member of the Research Committee of Biosphere 2 Center, Columbia University. Dr. Smith has authored several books and chapters addressing the ecology of the plants of the deserts of the Southwest.

Stanley Smith
Department of Biological Sciences
University of Nevada, Las Vegas
4505 Maryland Parkway
Las Vegas, NV 89154-4004
Email: ssmith@ccmail.nevada.edu

Preface

This book is a first attempt to look at a number of important Nevada environmental issues in a college textbook. It is a collaborative effort of many present or former faculty members of the University and Community College System of Nevada, with the Community College of Southern Nevada, the University of Nevada, Las Vegas, the Desert Research Institute, and the University of Nevada, Reno. While most of the authors have lived in southern Nevada for many years, we think that students throughout the state will find much of relevance in these pages, as we do not restrict our discussions to southern Nevada. Students from neighboring states may also find much to interest them here, and many of the issues that we face are the same ones that they face.

We have made every effort to make the book readable and accurate. Our narratives are a result of our experience in lecture and discussion with our students, and it is our hope that this approach reaches readers. This volume is a first edition in what will eventually become a more complete introduction to the major environmental issues in Nevada. We welcome viewpoint essays, which we would like to include in future editions. For the most part, our statements of opinion are kept to a minimum here. Where our opinions do appear, they are identified, but mainly we ask many questions and want readers to develop their own opinions. We simply put the background and the issues on the table to inform the student of the nature of environmental problems in general and in Nevada in particular. The order of chapters was chosen to match the order of largely distributed textbooks in environmental science; so this book can serve as a companion volume for an introduction to environmental science course.

The opening chapter, *Nevada's Environment: What, Me Worry?*, gives the reader the scope and history of Nevada environmental issues. The subsequent chapters introduce several important environmental issues facing the state, namely *Population Extremes*, *Nevada Land Management*, *Environmental Health, Air Pollution*, and *Nevada Energy Futures*. Another chapter, *Nevada Agriculture*, provides an overview of the state's third most important industry, one that requires a healthy environment. Other chapters provide a natural history context for the issues: *Nevada Biomes* and *Nevada Geology*. The final chapter, *Global Climate Change Research in Nevada*, is an in-depth look at an enormous collaborative effort to test climate change hypotheses in the Nevada desert as part of a global research program.

In the waning hours before my manuscript deadline, I am reminded of the many reasons why the papers that my beloved students have given me over the years are late or not good enough. Just as I can understand, but not accept, these reasons, I also cannot accept any of the reasons that so many mistakes remain in this manuscript. I can only accept responsibility for them. There will be some errors in content. When you detect them, please bring these to my attention as they are my fault alone. I think I have successfully kept these few in number and minor in consequence.

Acknowledgments

Why is "Acknowledgments" the last section I write, when the contributions of others are always so important in my own achievements? Once again, I am too late in beginning an accounting of other's assistance in this project. I would like to recognize all staff and chairpersons at the science departments of the Community College of Southern Nevada, University of Nevada, Las Vegas Biological Sciences Department, and also at the University of Nevada, Reno, Biological Resources Research Center. The contributions of all were essential. I believe I can freely speak for all the authors when I say they thank their families and loved ones for their patience.

Chapter 1

Nevada's Environment: What, Me Worry?

David A. Charlet

Figure 1.1 Atomic test "Grable" at the Nevada Test Site. U.S. Department of Energy Photograph.

Nevada Contributions to the Nation, and the National Perception of Nevada

How long have you lived in Nevada? Most residents have lived here less than five years. The common perception of newcomers and visitors is that the state is a desert wasteland. Indeed, the national perception of Nevada is that it is suitable for activities that would be unwelcome in their state.

Throughout its history, Nevada has sacrificed its biologic environment in order to extract its mineral wealth. Indeed, from our nickname ("The Silver State"), our motto ("All for our Country"), and our slogan ("Battle Born"), an ongoing and significant contribution of Nevada to our nation is clear. Sealing its admission as a state during the Civil War, Nevada's gold and silver were essential to finance the war effort of the Union. Formidable problems due to environmental factors such as topography and climate were overcome by many ingenious technological innovations in mining, dewatering, logging, and provision of domestic water supply. These innovations were stimulated by a bounty in the environment—the gold and silver that could be extracted by judicious use of other elements of the environment, the water, the forests, the grasslands, and forage in the mountains. Altogether, events leading to the admission of Nevada were not only remarkable, but they had everything to do with Nevada's environment: Nevada's mineral wealth was extracted by using the biologic wealth of its forests and the power of its waters. The wealth extracted was sufficient to allow the Union to win the Civil War, and left in its wake a prosperous human economy—these are remarkable achievements of both humans and Nevada's environment. This powerful interaction between Nevadans, the environment of their state, and the national interest can be followed to the present day.

In the early 1900s, Nevada was the site of the nation's first federal desert reclamation project that proclaimed to "make the desert bloom," the Newlands Project in Fallon, Nevada. Here the overflow waters of Lake Tahoe gathered in the Truckee River, flowing through Reno and to rest in Pyramid Lake were diverted to the Lahontan Valley, in which the town of Fallon and its beautiful farms grew. In the 1930s, Nevada played a crucial role in a massive federal water reclamation and energy project, Boulder (Hoover) Dam and Lake Mead. The project supplies water to Arizona, California, and Nevada, and electrical power to Arizona and California. During World War II, tungsten, copper, and titanium were in short supply but present in Nevada mountains, and the federal government helped support exploration efforts. Large modern copper mines near Yerington and Ely, Nevada, contributed significant copper for the war effort, and a titanium plant in Henderson continues to operate. During the Cold War, Nevada is where we detonated more than 1,000 nuclear explosives in aboveground tests from 1951 until 1963 (Fig 1.1), and below ground critical tests continued until 1992. In the 1970s, Nevada was selected as the site for the Cold War shell game known as the MX Missile System, in which it was planned that many Nevada valleys would house complex mazes of missile silos, a proposal eventually dropped. In the 1980s, the nation decided that Nevada was the suitable place to put the nation's nuclear waste. With solid waste sites rapidly disappearing in California, Nevada was seen as a perfect place to dump Californian trash. At the beginning of the 21st century, the nation not only continues to see Nevada as the nation's nuclear waste repository and solid waste dump, but also as an energy-generating producer for our neighbors: Arizona, Utah, and California. Many wind, natural gas, and biomass proposals for Nevada continue to be considered.

However, there is a lot more to Nevada than a convenient place to bury waste or to build power plants for other states whose residents do not want to have them nearby. For instance, Nevada's Humboldt-Toiyabe National Forest (NF) and the Desert National Wildlife Refuge (NWR) northwest of Las Vegas are the largest NF and NWR in the USA outside of Alaska. This textbook will help you to learn more about a part of Nevada you may not have known existed even though it is important to you.

Economic Importance of the Nevada Environment

It is interesting to examine the state's major industries and see how important the environment is to each industry. Tourism and Gaming is the most important industry in Nevada. While much of the attraction Nevada has to visitors is for the artificial environments it has created in Reno and Las Vegas, much tourism, especially international tourism, is also drawn to the area because of the natural beauty in and near the state. Las Vegas may have more National Parks, Monuments, and Recreation Areas (Bryce Canyon NP, Death Valley NP, Zion NP, Great Basin NP, Grand Canyon NP, Lake Mead NRA, Mojave National Preserve), within a half-day's drive than any other city in the USA. Even the state maps promote another side of Nevada by showing an ace card drawn, and the other side of the card is a view of Lamoille Canyon in the Ruby Mountains near Elko. While an environment filled with natural beauty attracts visitors, an environment that has the potential for accidents involving transportation of nuclear waste may keep visitors away. This is a concern for local and state community leaders.

Mining is the number two industry in Nevada, and we are not talking about "chump change." The total mineral production in the state was estimated at $2.94 billion in 1999, down 15% from its all-time high of $3.45 billion in 1996. The USA is the world's second leading gold producer, with Nevada producing more than 8 million troy ounces (troy ounce = 480 grains, 31.104 grams, or 1.0968 ounce), fully 75% of the nation's total. In fact, if Nevada were a nation, it would be the third greatest producer of gold. Moreover, with nearly 19.5 million troy ounces produced worth $103 million yearly, Nevada is the state with the greatest silver production in the USA. While gold and silver are still being extracted from Nevada mountains, environmental regulations must be met for the use of water and hazardous materials used in the processing of gold ore, and the toxic wastes released in the environment must be mitigated. New explorations and development require biological surveys, but always lead to habitat destruction and subsequent restoration. Indeed, the environment is very important to the mining industry, and there are many opportunities for creative environmental scientists to devise solutions to the environmental problems caused by mining.

Agriculture remains the third largest industry in the state, and its success in both farming and ranching is inextricably wound into the health of the environment. Successful agriculture needs excellent soils, and these soils must remain in good condition, if not improve. Adequate and dependable water supplies are essential to Nevada farmers, and so good vegetation conditions in the watersheds are vitally important to the farmers' well being, as well as conservation of groundwater and surface water resources. If you don't grow your own food, and if you think that's a farmer's problem and not yours, then how would you eat if someone else didn't grow that food for you? The farmer's problems are everyone's problems; and farmers are utterly dependent upon a healthy environment (and a lot of hard work).

Nevada's economy continues to boom, in spite of national economic downturns. Most of the gains have been achieved in the area of Tourism and Gaming, accompanied by a tremendous construction industry. It has been a monumental task to build enough homes and businesses to support the exploding population centers, mainly in the Reno-Carson and Las Vegas Valley areas, and the developers are meeting the demand. This growth has relied on the availability of large expanses of cheap land, seen rightfully by real estate developers, local government, businesses, and the citizens at large as economic opportunities. Ecologists acknowledge this too, but they also see the "vacant land" as **habitat**, a teeming, turbulent, living mass of nutrient cycling and energy transfer; reproduction rates and death rates; predation, parasitism, competition, and mutualism. Ecologists see in this empty land the latest snapshot in a three billion year timeline: the selection of the most fit in the struggle for life. It is a sobering viewpoint.

Ecologic Importance of Nevada

Nevada occupies a crucial place in western North America. Most of Nevada is in a region defined in various ways by different disciplines, but called by all the Great Basin. Wedged between the Sierra Nevada to the west and the Rocky Mountains to the east, the Great Basin is a high, arid environment hosting two of North America's deserts: the Great Basin desert and the Mojave Desert. Scattered throughout this arid region are wetlands at all elevations and high mountain forests. The wetlands are not only essential for the local wildlife, they also allow migratory birds moving along the Pacific Flyway from the Arctic to the Antarctic to rest and take water and food.

Throughout the region, plants and animals have evolved unique adaptations in response to the rigors of their environment. Nevada has more than 300 **endemic** species, or species that grow in Nevada and nowhere else in the world. Both the endemic and the common species operate together in interfingering ecosystems, themselves remarkably resilient and filled with natural wealth, producing food and lumber, clean air and fresh water.

It is no coincidence that ecology and economy begin with the same first three letters. *Eco-* comes from the Greek word *oikos*, and means "house." The *–logy* at the end is the same *-logy* we find at the end of many disciplines, geology, psychology, biology, and others, and comes from the Greek work *logos*, meaning "word," and which we have come to use in a way that means "study of." Therefore, "ecology" (*eco-* = house; *-logy* = study) literally means "house word" but we use it to mean the study of our environment. *Eco-* in economy also means house, but *–nomy* comes from the Greek word, *nemien*, and means "to manage." Thus, economy (*eco-* = house, *-nomy* = to manage) literally means, "to manage the house," but we use it to mean managing money. Ultimately, all money comes from the Earth; and ecosystems create most of that wealth. Even many of our geologic resources were created by ecosystems: oil fields are simply ancient, preserved microbial ecosystems that have rotted and cured. Coal deposits are ancient forest ecosystems. Both coal and oil have so much life energy left in them, we can burn them and power modern industry. The sedimentary rock limestone was created from the exoskeletons of sea creatures, crushed and subjected to enormous pressures. When pressures are great enough and other conditions are just right, the limestone can become the metamorphic rock, marble.

When ecosystems begin to lose their stability, their bounty becomes undependable, and the unpredictable adjustments can be temporarily destructive. These adjustments are particularly disruptive to human commerce and industry. As forests throughout western North America adjust to changes in weather, fire regime, and forestry, they burn. These fires cost US taxpayers billions of dollars to fight, and more billions of dollars worth of standing timber lost to industry. Whether we are concerned with wild animals and plants or not, whether we farm the land or not, whether we earn or living by mining, timber, ranching or any other means, if you live and work in Nevada, then Nevada's environment is very important to you. It is in your best interest to learn what you can about it, and for Nevada's ecosystems to be stable as long as you remain in the Silver State.

Chapter 3

Nevada Biomes

David A. Charlet

Terrestrial Biomes of the World

Levels of Organization: Ecosystems, Landscapes, Biomes

What defines a **biome**, and how do we know which one we live in? Technically speaking, a biome is a large region of the earth that includes all living organisms as well as the non-living components of rock, air, and water. When thinking of what a biome is, it is useful to consider it as another classification the organization of nature, at a level larger than an **ecosystem** or a **landscape**. This is the ascending order of scale of these levels of organization: ecosystem, landscape, and biome. In practice, the word "ecosystem" is used to describe an interacting biota along with its abiotic components at several levels of scale, often including the landscape scale. However, if we want to be very precise, it is best to limit the use of the word "ecosystem" to small areas, generally small enough for us to see its boundaries. For instance, if we consider a pond and all of its organisms, this is an ecosystem. Its boundary is clear: the pond ends and the terrestrial ecosystem surrounding it begins. Organisms inside the pond interact much more with one another than they do with their terrestrial counterparts outside the pond. This is not to say that the pond has no connection with the surrounding terrestrial ecosystem, that there are no interactions between organisms in the pond or on the ground, or that there is no nutrient flow between the two ecosystems. But it is easy to draw a line around the pond, and easy for us to see that the organisms in the pond are different from those outside the pond. Therefore, for our convenience, so we can understand the dynamic interactions between organisms of the pond with each other and with the pond itself, we say that the pond is an ecosystem.

The landscape scale of organization contains many interacting ecosystems. The best way to think about this scale is simply to look on a map for names of large landscape features, such as "Spring Mountains" and "Ruby Valley." Any named mountain range, plain, valley, sand dune complex, or large canyon can be considered the "landscape" level of organization. In a mountain range, for instance, numerous plant communities are clearly different from one another, even to someone who knows no plant species. Often how we tell that we have left one plant community and have entered another is not so much from a change in species, but in a change of vegetation **structure**, a topic we will explore in some detail in the next section. For now, suffice it to say that on a mountain you have ponds, meadows, forests, shrublands, and creeks. Each of these units is easy even for the amateur to recognize (although more difficult to draw lines around and name), and each can be considered to support communities (a group of organisms interacting with one another at the same time and place). If we include the abiotic factors with which these organisms are interacting, then we are thinking about ecosystems. If we consider a named landscape feature like a mountain range, and all of the ecosystems occurring on that landscape feature, then we are thinking about the landscape scale of organization. Again, it is easy to see where the mountain is, but it is more difficult to say exactly where it ends and the adjacent valley begins, a problem with which landscape ecologists and geologists still struggle.

What, then, is the scale of a biome? We are told a biome is "a large region of the earth," yet this is vague — how big is the region are we talking about? How can you tell where one biome ends and another begins? Since a biome is a larger level of organization than a landscape feature, we should expect that biomes would be larger than landscape features. Usually they are, but they do not have to be. Sometimes a single landscape feature actually has pieces of many biomes on it simultaneously. So, we need to explore this topic in detail.

It is perhaps best to think of using biomes as a method to separate all terrestrial environments in the world, dividing them into a few classes or groups. Different workers have tackled this exercise with different classification schemes, and consequently different classes and maps. Some workers have divided the world into as few as ten major biomes, while others define more than 20. As a result, when one first encounters the concept and sees the many different classification schemes and names for the world's biomes, it seems that there is no agreement. Regardless of the individual worker's methods and final map and names, the methods employed for recognizing, classifying, and naming biomes are essentially the same. Further, the environmental factors that control the distribution of biomes are identical. While there is no consensus on exactly how to cut the pie of the world's terrestrial systems, there is agreement that there is a level of organization we can call a biome. Biomes are generally much larger than landscape units are. Even though pieces of the same biome do not necessarily have the same species, the species that are in both pieces use similar strategies to live in a similar environment.

Recognition of Biomes

Although biomes include much more than just the plants in an area, biomes are recognized by the sum total of all plants an area, or its **vegetation**. This may seem contradictory, because earlier when defining a biome, we said that a biome includes all organisms and their abiotic environment. Yet, we use only the plants to recognize and name a particular biome. Why do you suppose this is so?

Moreover, for the purposes of recognizing and naming biomes, biologists are not concerned with *what species* occur there, only *what kind of species* occur there. In fact, you do not need to be a botanist at all to recognize biomes. All you need is some common knowledge about plants (most of which you already possess), and a few rules to apply, and you will be capable of recognizing and naming any biome in which you find yourself.

If we are going to divide the whole world into 10-15 different "types" or classes of vegetation, then we have to think big. Examine the map of the world in your primary textbook— see how large most of the colored shapes or—**polygons** are. Look especially in an area you know well, and then find it on a highway map to get some idea of the enormous areas on the ground these areas represent. You must think big to understand biomes.

All of the critical features of biomes are represented in what for our purposes I call the "complete name" of the biome. This "complete name" contains all the descriptive elements that are used in part for all biomes. If we take all the names ever used for biomes and classify them as to what they attempt to describe, we find that the staggering number of terms takes on a logical order. One way to order these names is as follows:

> *World region, lifestyle, reproduction, leaf type, [life form], vegetation structure.*

I put "life form" in brackets because it is never used in the name of the biome, but as a means to determine and name the vegetation structure. What is meant by these terms? How do you recognize them when you see them?

World Region

World region refers to three main divisions of the world based on latitude. These areas are **Polar**, **Temperate**, and **Tropical**. If one of these terms is used to name a biome, it is always first (e.g., Tropical Forest). There are two Polar regions, one north of the Arctic Circle (66°30' N), and one south of the Antarctic Circle (66°30'S). There is one Tropical Region centered about the Equator, ranging north to the Tropic of Cancer, 23°27'N and south to the Tropic of Capricorn, 23°27'S. The two Temperate Regions occur in both the northern and southern hemispheres, between the Polar Region and the Tropical Region.

In the northern hemisphere, the Temperate Region occurs between 66°30'N and 22°27'N. We can also use the first word of a name to refer to a world region not based on latitude, but rather on elevation in a particular area. That term is **alpine**, and refers to areas in mountains that are too high to have a growing season long enough to support woody plants. Another name you will occasionally encounter in this position is "**boreal**." This is part of a paired set of terms referring to the northern and southern regions of the world. Boreal means north, and we recognize that it is the Latin name for the northern lights, Aurora Borealis. **Australis** means south, and we recognize it in the name Australia and of the southern lights, the Aurora Australis.

Vegetation Structure

Next, let us deal with vegetation structure, usually used to provide the last word in the name of a biome. In order to know vegetation structure, we need to first determine the **life form** of the **dominant plant** in the area. For the purpose of recognizing and naming biomes, by dominant plant we mean the plant life form that possesses the most **biomass** in the area. By life form, we mean the general form or "type" of plant. Relevant here is that there are only four life forms you need to be aware of, because at this time only these four life forms of plants dominate entire biomes. To discover the life form of a plant, he first thing you ask yourself is, "Is the plant woody or not woody?" Wood is simply the dead, hardened cells of a plant stem grown in an earlier year. The fact that a plant has wood is very significant. It tells us that the plant stores above ground its buds, or masses of tissue ready to elongate and develop into additional stems, foliage, and/or reproductive structures. Woody plants that occur in temperate climates must protect their buds during cold winter temperatures by hiding them inside the wood. Non-woody plants use a different strategy to solve the same problem.

We can separate woody plants into two major types according to their form, trees and shrubs. Trees usually have upright stems that branch well above the ground. As a rule, you can say that a tree usually branches above the height of a person 1.8m (ca. 6ft) tall. On the other hand, shrubs usually branch profusely around their base, right at or just above the ground. Lots of names are used to describe shrubs, such as chaparral, scrub, bush, and brush, but all of these terms refer to a low woody plant with many branches close to the ground. We will always call such plants shrubs in this book.

The other two plant life forms either store their buds underground or do not store their buds at all. These life forms have no wood, and so are called non-woody plants. They come in two general forms, **herbs** and **graminoids**. Herbs are plants with relatively broad leaves, conspicuous flowers, and netlike veins, whereas graminoids (*gramin-* = grass, *-oid* = -like; grass-like) are grass-like plants usually with linear leaves, inconspicuous flowers, and parallel veins.

And so we now know the four life forms of plants abundant enough to dominate large regions of the world: trees, shrubs, herbs, and grasses and grass-like plants (graminoids). When grasses dominate the land, we call the formation a **grassland**. Large grasslands occur in North and South America, in Africa, and in Eurasia. Shrubs dominate large tracts of the earth also, and shrub-dominated areas are called simply **shrublands**. When trees dominate a large area, and the canopy created by trees covers about 50% or more of the land beneath them, we say that this is a **forest**. We could also say that we are referring to this community as having the vegetation structure of a forest. If the dominant life forms are trees, but the trees are more open and scattered (less than 50% cover), then we refer to this vegetation structure as being a **woodland**. When herbs dominate small areas in the mountains, we call these **meadows**, but on the scale of biomes, we do not use that term. Instead, we use the word **tundra**, a Russian term used to refer to the meadows lacking woody plants North of the Arctic Circle and above the upper treelines in the mountains.

Now we have encountered most of the words used in the final place in the name of a biome: grassland, forest, woodland, and tundra. Like "meadow," the word "shrubland" is only rarely used as the final word in a biome's name. Instead, another name is used for biomes dominated by this lifeform: **desert**. Personally, I prefer to retain the word "shrubland" at the end of a biome's name. Nevertheless, by convention most people and textbooks use the word desert alone, applied to all kinds of deserts that can be more accurately described using this system. Still, in our arsenal of names now, we can come up with quite a few combinations that are used as names of biomes in textbooks: Alpine Tundra, Tropical Forest, Temperate Woodland, and Polar (Arctic) Tundra. But other words often appear in biome names, and they can help divide these very general types into more useful classes.

Some words are simply regional names for large formations that occur in a particular region. For instance, Prairie (North America), Savannah (Africa), and Steppe (Asia) are all regional names for Grassland. Taiga and Boreal Forest are both regional names for the same cold-tolerant, temperate evergreen needleleaf conifer forests of the far northern hemisphere.

Lifestyle

The **lifestyle** of a plant refers to its life-history strategy as it attempts to survive and reproduce. For woody plants, there are two lifestyle types, and two different lifestyle types exist in non-woody plants. In woody plants, the question is: Does the plant hold on to leaves throughout the year (**evergreen**)? Or, does it drop its leaves all at once (**deciduous**)? In fact, deciduous means losing leaves, and all plants lose their leaves, even evergreen plants. Did you know that? Think about the last time you walked through an evergreen forest, like a pine forest. What were you walking on? That's right, pine needles; and pine needles are the needlelike leaves of a pine tree. Even though most of the trees you were walking under were alive and had needles (they are evergreen), nevertheless you were walking on their leaves. All trees and shrubs lose their leaves; it is simply that some trees and shrubs keep leaves on themselves all year so that whenever we look at them they are green, or "evergreen." Other trees and shrubs lose their leaves all at once, either in winter when it is cold (**seasonally deciduous**) or when it is too dry (**drought deciduous**). We use the terms evergreen or deciduous only to describe the woody life forms of trees and shrubs. The lifestyle "choice" for a woody plant to be deciduous or evergreen is highly significant. Can you think of advantages and disadvantages of each of these strategies? In situations where the advantages outweigh the disadvantages for a particular lifestyle, we see those kinds of plants dominate. When the conditions are the opposite, we see the plants "making" the other lifestyle choice dominate. Commonly, all lifestyles can be found in the same community, and isn't that interesting? What do you think about that observation?

Non-woody plants also have two lifestyle strategies, but what we are concerned with here is whether the plant lives out its entire life in one growing season (**annual**) or stores its buds underground (**perennial**). As with the evergreen/deciduous dichotomy, the different lifestyle strategies succeed in different situations. Annual herbs and graminoids are most successful in arid areas with undependable and infrequent precipitation, and they never dominate large enough areas to use the name "annual" for any biome. However, in our Mojave Desert, like most deserts throughout the world, the annual lifestyle succeeds because the plant avoids dry periods by only sprouting when there is sufficient moisture to complete a reproductive cycle. Then they "store" their offspring as seed, ready to sprout only when there is sufficient moisture available to complete their lifecycle. While annuals do well in desert environments, they rarely actually dominate whole deserts. The domination of deserts is usually left to shrubs, which are well-suited to develop deep-root systems that can reach more dependable water sources deep in the ground, and whose small size tends to reduce water needs and losses. So, desert biomes are usually dominated by shrubs, but these shrubs are accompanied by a wide variety of annual plants that show themselves perhaps as rarely as once a decade. Herbs of any lifestyle dominate enough area on earth to be called a biome only in the tundra, but here they do so for a very different reason. Thus the kind of herb that dominates has made a different lifestyle choice. Here, the growing season is seldom long enough for an annual plant to complete an entire life cycle in one season, and yet the temperatures outside the growing season are so severe that annual plants cannot survive the frost and snow. Further, here woody plants do not have enough time to "harden off" and protect their buds from the bitter cold above the ground. In the tundra, the perennial lifestyle is the most successful. Perennials will grow when they can, but when conditions get too cold, they hide next year's buds underground in a nourishing rootstock. It may be many years before the season is long enough for the plant to put forth a reproductive effort, but in the intervening years, it will store accumulated photosynthetic product in its root, and wait for a suitable time to try to reproduce sexually. In the meantime, many tundra perennials can reproduce vegetatively (asexual reproduction), by runners and underground stems. Similarly, annual grasses or grass-like plants almost never dominate the great grasslands of the world, but instead these are dominated by perennial graminoids. When naming biomes, we seldom if ever need ascertain the annual or perennial lifestyle of the dominant plants, because in most cases they will be perennial. Instead, it is enough to say that the biome is tundra or grassland.

Reproduction

When referring to the "reproduction" of the dominant species, we only need to concern ourselves with this point when considering biomes dominated by woody plants. By "reproduction" what I mean is the type of reproductive structure. There are many types of reproductive structures on plants, but the only kinds of relevance here (in that they can dominate a biome) are cones and flowers. That is, woody plants capable of dominating a biome have sexual activities (make gametes and the gametic nuclei join to make a new individual) in either cones or flowers. There are more technical botanical definitions, but in general, you can recognize a cone-bearing plant (**conifer**) as having persisting woody structures (**cones**) containing ovaries that will become the seeds (e.g., "pine nuts"). **Flowers** are soft, ephemeral, and often colorful and aromatic, with ovaries that develop into some kind of woody (maple seed wing) or juicy "fruit" (the cherry) containing a seed (the "almond" inside the pit). If a plant is woody and dominates a biome, it either is a conifer (bearing its reproductive structures in cones), or a flowering plant (bearing its reproductive structures in flowers).

Leaf Type

For biomes dominated by woody plants, we need a number of descriptive terms that further define the type of woody plants we find there. **Leaf type** is the last such term we will consider. Here we see two kinds whose names appear in biomes: **broadleaf** and **needleleaf**. Leaves are the primary organ of photosynthesis for most plants, and so are responsible for the plant's food production. The shape of the leaf represents another choice in the life strategy for a plant species, as broadleaves are advantageous in certain situations while needleleaves have other advantages. Therefore, the shape of the leaf gives us clues about a plant's living conditions. If a plant has broadleaves, it can absorb much light and grow very fast, but it needs a tremendous amount of water to support it. Broadleaf plants are common in areas of low light (forest floors) or high water (tropical rainforests). Needleleaves are not good at capturing light, but are excellent in conserving water, so we see needleleaf plants succeeding in conditions of high light and low temperatures and/or precipitation; areas like deserts, arid conifer woodlands, boreal forest. Sometimes in place of leaf type, workers will use the name of another prominent feature of a tree or shrub, the presence of thorns for instance, in the name. Landscapes in Arizona that are dominated by Saguaro cactus should be called, ***"Temperate deciduous flowering thorn woodland***, " because Saguaro leaves are ephemeral and associated with thorns. Saguaros are tall plants with woody stems, usually branching above 1.8m, and so are trees. Because the trees are widely spaced, the formation is a woodland, not a forest.

Putting It All Together

Now you know all the terms that you will need to recognize a biome and construct for it a complete name (Table 3.1). The first step in recognizing a biome is to determine the vegetation structure. What kinds of plants are holding the most biomass, woody or not woody? If woody, are they trees or shrubs (forests, woodlands, and shrublands)? If they are trees, is the canopy nearly closed (forest) or is it very open (woodland)? If not woody plants, are they herbs (tundra) or grass-like (grassland) plants? Now you know the last word of your biome: forest, woodland, grassland, or if the dominant life form is shrubs, we call this shrubland a desert. For forests or woodlands, we have other terms we can use to further describe them. We need to decide whether they are evergreen or deciduous, coniferous or flowering, and needleleaf or broadleaf. Once you have made these determinations, you are ready to fill in the names in the correct order to name your biome. Naturally, you need to know what part of the world you are considering (Polar, Temperate, or Tropical). The exception here is that you might be on top of some giant mountain somewhere above the highest trees, in which you case you are in the Alpine region. Imagine yourself sitting up at the South Loop Mount Charleston trailhead parking lot and asking yourself, "In which biome am I?" Make yourself a table of 2 rows with 6 columns, using the words world region, lifestyle, reproduction, leaf type, [life form], and vegetation structure across the top row of cells. Wherever you are in Nevada, you are between about 34°30'N and 42°N, which is the Temperate world region, so fill in the empty box below World Region with "Temperate."

World Region	Lifestyle	Reproduction	Leaf Type	[Life Form]	Vegetation Structure
Temperate					

As you look around, you see that most of the biomass is in woody plants that are upright with single straight stems. You know plants with these characteristics are trees, so that is the dominant life form, and you fill in the empty life form cell with "[tree]." You can put it in brackets and lower case to remind yourself that you won't actually use that word in the name for the biome. You need this in order to name the vegetation structure. You know that when the dominant life form is trees, they can make either forests or woodlands. You see that the forest here has a thick canopy and so you know you are in forest, not woodland. Fill in "forest" in the vegetation structure cell.

World Region	Lifestyle	Reproduction	Leaf Type	[Life Form]	Vegetation Structure
Temperate				[Tree]	Forest

What can you say about the trees that dominate the forest? For lifestyle, you need to answer the questions: are these trees that lose their leaves (deciduous)? Are they trees that stay green all year (evergreen)? Or, are there so many of both it is difficult to determine which has most of the biomass (mixed)? After inspection, you will find that while there are some of both types, by far most of the trees are evergreen. To fill in the reproduction box, you determine whether the dominant trees reproduce by cones or in flowers, and you will find that most of the evergreen trees there are reproducing with cones, and so you write conifer in the reproduction box. Finally, you examine the leaves of the dominant trees and determine that they are like needles and not at all broad. This means that you have filled in your box and correctly determined the biome that occurs at this spot in the world, saying that it is a ***Temperate evergreen conifer needleleaf forest***. In so saying, you have described concisely and accurately the conditions in which you are located for someone in another part of the world. No matter where you find yourself, you can ask the same questions and name the biome.

World Region	Lifestyle	Reproduction	Leaf Type	[Life Form]	Vegetation Structure
Temperate	Evergreen	Conifer	Needleleaf	[Tree]	Forest

Take care not to assume that just because conifers dominate the forest that it needs to be both evergreen and needleleaf, or that because a shrubland is dominated by flowering plants that it must be broadleaf and deciduous. For instance, rosemary is an evergreen shrub with needle-like leaves, but it has dainty purple flowers. If this plant dominates large parts of the hills of its native Greece, then this would be ***Temperate evergreen flowering needleleaf shrubland desert***. A small portion of central China is dominated by a seasonally deciduous conifer, the dawn redwood (*Metasequoia glyptostroboides*) related to the coast redwood (*Sequoia sempervirens*) and giant sequoia (*Sequioadendron giganteum*) of California. A forest dominated by this tree could be called a ***Temperate deciduous conifer needleleaf forest***. Many flowering trees have broad leaves that are deciduous, but some such as magnolia (*Magnolia*), Pacific madrone (*Arbutus* sp. [sp. = one species]) and some oaks (*Quercus* spp. [spp. = more than one species]) are evergreen. The ginkgo (*Ginkgo biloba*) is a primitive gymnosperm, much more closely related to conifers than flowering trees. Yet, the ginkgo has broadleaves that it discards every autumn. Larches (*Larix* spp.) are needleleaf conifers of the high mountains and northern latitudes with a seasonally deciduous lifestyle. It is fascinating to see the large number of strategies being employed by an assemblage of plants in any particular area.

Table 3.1. Vegetation: The Key to Recognizing and Naming Biomes. Combine terms in the table in order to make a name for a biome. Choose one term from each column except for life form column (bracketed [ ]) and final column. The life form column is used to determine the vegetation structure. If dominated by trees, then it is forest; if dominated by grass, then it is grassland. Terms: graminoid—grasses and grass-likes plants with linear leaves and inconspicuous flowers such as sedges, rushes, tules. Woody species—plants with wood, buds overwintering above ground. Shrubs and trees (cactus are either shrubs or trees). Annual—entire life cycle in < 1 yr: graminoids and herbs only. Perennial: buds overwinter below ground; graminoids and herbs only. Herbs: broadleaf, non-woody plants, usually with conspicuous, ephemeral flowers.

World Region	Lifestyle	Reproduction	Leaf Type	[Life Form]	Vegetation Structure Where Lifeform Is Dominant	Biome Where Lifeform Is Dominant
Polar	Evergreen	Conifer	Needleleaf	[Tree]	[Tree] Forest	Any Forest Any Woodland
Temperate	Deciduous	Flowering	Broadleaf	[Shrub]	[Shrub] Shrubland	Desert
Tropical	Annual			[Herb]	[Herb] Tundra / Meadow	Tundra (Alpine & Polar)
Alpine	Perennial			[Graminoid]	[Graminoid] Grassland	Any Grassland Prairie Savannah Steppe

Control of Biome Distribution

Most textbooks maintain that the topography and climate of a region control the distribution of biomes. While topography interacts with climate to create conditions suitable for the survival and reproductive success of suites of plant species with particular life history and morphological strategies, we cannot look at topography alone to predict what kinds of species will occur in an area. However, we can look at only two simple measures of a climate (annual average temperature and average annual precipitation) and successfully make such predictions. So while it is correct to say that topography plays a role in the formation of a biome, it is not the critical factor: climate is.

For instance, the Tundra biome occurs in two types of situations with dramatically different topography: Alpine and Polar. Alpine Tundra occurs in some of the world's most severe and tortured topography: the tops of the highest mountains in the world. However, Arctic Tundra occurs in the plains of the Arctic. Few places on earth rival the monotony of this region's flat, featureless landscape. In North America, the Grassland biomes occur in a wide variety of topographical situations, on plains in the Great Plains, on hills in the Palouse Prairie of Eastern Washington, as well as in small pieces on mountain ranges in Central Nevada. Tropical rainforest occurs on plains, in valleys, and in mountains.

Biomes of Nevada

Once told of biomes, most Nevadans are usually shocked to discover that most of the world's major biomes are here in their home state. I imagine you will be surprised to learn that Nevada possesses seven of the world's ten major biome divisions as presented by Whittaker's figure in your primary textbook.

The most dramatic example of this is just west of Las Vegas, in a walk from the Amargosa Valley east to the top of Mount Charleston. One hundred fifty years ago, a walk through as many biomes in about half the distance would have been possible, but one of the biomes no longer exists in the Las Vegas Valley. Now we must now go to the Amargosa Valley to find the biome. So get your maps out, and imagine yourself in the Ash Meadows National Wildlife Refuge west of Pahrump, in the Amargosa Valley of Nye County, Nevada.

A small set of northwest-trending hills separates Pahrump Valley from the Amargosa Valley. This range of hills is called "Devils Hole Hills" because of a curious feature at the southeast base of the northernmost hills in the range, Devils Hole. Devils Hole is a unique pond wedged between two limestone towers. Its tiny opening, only a few square meters that are exposed to the air, opens into enormous caverns of unknown depth beneath. The small area that opens to the air at the bottom of the hole has a shelf less than 2m deep in which is a unique **freshwater ecosystem**. Here in this little pond are all the individuals of the Devils Hole Pupfish (*Cyprinodon diabolis*). This is not the only fish or spring in the area. This is only one pond, in a curvilinear feature of occasional springs that spans the entire western front of the Devils Hole Hills, a length of about 30 km. It has a collective discharge of about 30,000 gallons per minute. This incredible amount of water supports the formation of Ash Meadows' **wetlands**, large grasslands, forests, and woodlands. It is here that we will begin our walk. We can call these plant communities "***Temperate perennial grassland***", "***Temperate deciduous thorn forest***," and "***Temperate seasonally deciduous thorn woodland***." The communities represent patches of the biome for which we have named them, and thus give us an idea about how natural areas of similar condition throughout the world would appear, even though the species are probably all different. Outside the area of the springs and their overflow, we enter a shrubland of many different species, but usually dominated by a tall shrub with small, oily green leaves and bright yellow flowers. This evergreen shrub is the creosote bush (*Larrea tridentata*), and this shrub-dominated community is a desert we can call a "***Temperate microphyllous*** (small-leaved) ***evergreen shrubland desert***." As we walk through the northern end of the Pahrump Valley on our way to Mount Charleston in the Spring Mountains, we will pass through many different kinds of desert shrublands, well into the mountains themselves. At around 4,500 feet elevation or so, we will start encountering trees as the major lifeform. At their lowest elevations, the trees are very open, and so we call this a woodland. The low elevation trees are usually yuccas, either the taller Joshua tree (*Yucca brevifolia*) or Mojave yucca (*Yucca schidigera*). Woodlands dominated by either of these species would be called, "***Tem-***

perate evergreen broadleaf flowering thorn woodland*,** " as the leaves have spines on the ends of them and persist all year on the tree. Higher on the mountain, trees get closer together. Once they do they are accompanied by short evergreen needleleaf conifers, pinyon pine (*Pinus monophylla*) and Utah juniper (*Juniperus osteosperma*). By 6,500 ft or so, they so dominate the landscape one can call it "Temperate evergreen needleleaf conifer woodland***." At around 7,000-7,500 ft, you will start to encounter much larger evergreen needleleaf trees, tall pines (*Pinus* spp.) and fir (*Abies* spp.), so thick that the canopy is at least half closed. You are now in a "***Temperate evergreen needleleaf conifer forest***." You will notice patches of another kind of tree here, with white stems and broad leaves that get bright yellow before they drop in the fall. These trees are aspen (*Populus tremuloides*), and their patches represent a "***Temperate deciduous broadleaf forest***." Higher still, from about 9,500 to 11,000 ft, you are in a evergreen needleleaf forest of extreme conditions. It is at the upper elevation limits of the tolerance of trees, a zone some call the "**subalpine**." This refers to its location at treeline, immediately below (sub) the alpine. These subalpine forests can be referred to as being part of the "***Boreal evergreen needleleaf conifer forest***" biome. Once you finally step above this forest, you are now in a landscape dominated by perennial herbs, and so you have reached the ***Alpine tundra*** near the top of Mount Charleston. On this walk the general biomes we have walked through include: Alpine Tundra, Temperate Coniferous Forest, Temperate Broadleaf Forest, Temperate Thorn Forest, Temperate Grassland, and Temperate Desert. We also encountered Freshwater Ecosystems and Wetlands. Can you draw a map of your journey on the figure of biomes and annual precipitation and temperature by Whittaker? If so, what does that tell you about why you saw what you saw where you saw it?

Chapter 4

Nevada Land Management and Conservation

David A. Charlet

Nevada Land: Overlaying Mosaics

Land Management Mosaic: Who Owns Nevada?

A contentious issue for many longtime Nevada residents is the management of lands in Nevada by the federal government. About 85% of Nevada, more than any other of the lower 48 states, is managed by federal agencies. But the federal government is not the only government that manages land in Nevada: state, county, and local governments also manage land in the state, leaving about 10 million acres, or less than 15%, as private property. About 5 million acres were given by the federal government to the Central Pacific Railroad in the 1860s in return for building a portion of the Trans-Continental Railroad. More than 1.5 million acres of this land is still owned by Southern Pacific Railroad, the single largest private landholder in the state.

Court cases concerning a few ranchers who allow cattle to graze without paying for permits or who make unauthorized changes to public land continue to appear in the news. These ranchers and their supporters insist that the federal agencies have no authority to grant permits and seize their cattle. Nye County Commissioner Dick Carver found himself on the cover of *Time* magazine for using a bulldozer to challenge federal employees and open the Jefferson Summit Road in 1994, a road that the USDA Forest Service had closed. The US government took Nye County to court and exerted its federal authority over Nevada public lands in a landmark case settled in 1996. One of the points that Nye County made was that the US Government did not have title to public lands in Nevada. The Federal District Court saw this otherwise, claiming the federal government had title to, and authority over, the land.

The Nye County case greatly interested the Western Shoshone. The Western Shoshone is a Native American tribe native to Nevada and they and their ancestors occupied most of Nevada continuously for at least 3,000 years. Some Western Shoshone also have been fined and had cattle removed for permit violations. The Western Shoshone contend that their 1863 Treaty of Ruby Valley, ratified by the US Congress and signed by the President, acknowledges that the disputed land (most of Nevada and parts of California and Idaho) belongs to the Western Shoshone. Further, treaties are made between two sovereign nations, and so the fact that the treaty exists acknowledges that the Western Shoshone people are a nation. The federal government attempted to settle this and other Native American land disputes throughout the US with the Indian Claims Commission Act of 1946. After years of negotiations and court cases, the Commission was disbanded in 1978, and the Western Shoshone case was transferred to the US Court of Claims. Finally, in 1979 the court set aside the amount of $26 million for the Western Shoshone to cease their claim and end their dispute. This amount was calculated by the value of the land in 1872. However, court cases continued through the 1980s to the present and to date the Western Shoshone have not taken the money set aside for them, for to do so they would give up their land claim. At publication time, US Senator Harry Reid (D, NV), was negotiating a settlement with all parties.

Federal Government

While it may still be unclear who owns Nevada (since there are three different interests each claiming ownership rights), it is clearer who controls Nevada ecosystems. The federal government is by far the entity with the largest (by area) area over which to exert authority over ecosystems. Fully 85% of the state is administered by the federal government of the United States, with ten agencies in four departments represented: Defense, Interior, Agriculture, and Energy (Table 4.1). About 40 million acres in Nevada are administered by the Bureau of Land Management, the largest single land manager in the state.

Table 4.1. Federal Government Land Managers in Nevada and Their Domains

Department	Agency	Acronym	Amount of Land (72,000,000 acres total)	Types of Management	Examples
Interior	Fish and Wildlife Service	USFWS	2,320,593 acres 3.2%	National Wildlife Refuges (NWR)	Stillwater NWR Pahranagat NWR Ash Meadows NWR Moapa Valley NWR Desert Game NWR Sheldon Antelope Range NWR Ruby Marshes NWR Fallon NWR
Interior	National Park Service	NPS	77,180 acres+ LMNRA+ DVNP	National Parks (NP) and Recreation Areas (NRA)	Great Basin NP Death Valley NP Lake Mead NRA
Interior	Bureau of Land Management	BLM	48,000,000 acres 66.7%	"Public land"	BLM lands Red Rock National Conservation Area
Interior	Bureau of Indian Affairs	BIA	1,152,000 acres 1.6%	Indian reservations Indian colonies	Western Shoshone (e.g., Yomba) N. Paiute (e.g., Pyramid Lake, Summit Lake) S. Paiute (Las Vegas) Washoe (Woodfords, CA/NV) Mojave (Fort Mojave)
Agriculture	Forest Service	USDA FS	6,000,000 acres 8.3%	National Forests	Humboldt-Toiyabe NF Inyo NF

Table 4.1. *Continued.*

Depart-ment	Agency	Acronym	Amount of Land 72,000,000 acres total	Types of Management	Examples
Defense	Army	US Army	N/a	Army bases	Hawthorne Weapons Storage Facility
Defense	Navy	US Navy	N/a	Naval bases	Fallon Naval Air Station
Defense	Air Force	USAF	N/a	Air bases	Nellis Air Force Base Nellis Air Force Range Las Vegas Gunnery Range
Defense	Coast Guard	USCG	N/a	Coast Guard bases	Lake Tahoe Coast Guard Station, Loran Facility, several others
Energy		DOE	3,567 km^2 881,049 acres (1.2%)		Nevada Test Site

The US Government is selling land under the Southern Nevada Public Land Management Act (Public Law 105-263). Under this Act, federal lands in Clark County in the immediate vicinity of Las Vegas are being disposed of in public auctions in order to provide opportunities for private enterprise and acquisition of environmentally sensitive lands for the protection of biodiversity. The proceeds are divided into several areas: 5% to the State of Nevada for the general education program, 10% to the Southern Nevada Water Authority, and 85% to a special account, the main purpose of which is to acquire environmentally significant lands.

Private Land

About 15% (about 10 million acres) of Nevada is owned by private parties or the state. Most of this private land is used for agriculture (see Chapter 5) and a considerable portion is administered by the state (see below) as parks, wildlife management areas, and prisons. Counties also maintain and operate parks, and all are wonderful. Clark County manages the largest park in the developed portion of the Las Vegas Valley, Sunset County Park. While much of the park has tennis and basketball courts, walk paths and lawns for people to enjoy, there is also a significant portion of the park south and east of the developed area that has natural sand dunes stabilized by honey mesquite (*Prosopis glandulosa*) and native shrubs.

State and Local Government

State government manages only a very small portion of Nevada's ecosystems. Several agencies have direct control ecosystems on land owned by the State of Nevada, most notably the Nevada Division of Wildlife (NDOW) and the Nevada Division of Parks (NDP). The Nevada Department of Transportation (NDOT) controls the ecosystems within the easements of state highways (not all of which are paved). Work details manned by prisoners under the Department of Prisons and directed by the Nevada Division of Forestry (NDOF) can often be seen applying herbicides, cleaning trash, pruning and landscaping, along Nevada's highways in green NDOF trucks. NDOW manages many Wildlife Management Areas (WMA) in the state, mostly in areas of abundant water with wetlands, marshes, riparian areas, springs and other lush vegetation, thus stopover points for birds using the Pacific Flyway. The Nevada Department of Parks manages land in the state's park system. If you have not gone to one of Nevada's many state parks (Table 4.2), you owe it to yourself to consider a journey to one a fieldtrip for yourself. The Park System provides information, maps, maintains trails, and most state parks have wonderfully maintained camping sites that you may use with a reservation and for a nominal fee.

Ecologic Mosaic

To most early non-native travelers, the trek through Nevada represented a necessary step to get to their real destination. The first European explorers in the state were the Spanish, who established the Spanish Trail through southern Nevada in the late 1700s. Jedidiah Strong Smith was the first white person known to cross the entire state, which he accomplished in 1827. The explorer Joseph Reddeford Walker crossed the state with a large party in 1833. The first mapping and scientific expeditions in the state were led by Colonel John C. Frémont from 1842 to 1846. It was not until 1849 that the first permanent settlement in Nevada was created at the site of modern Genoa, in Douglas County.

In 1854, E.G. Beckwith led a survey for the Pacific portion of the trans-continental railroad, but little was contributed from this survey concerning the biota of the state. After the discovery of gold and silver, however, more surveys were conducted in the state that included both geologists and biologists. The first attempts to catalogue the organisms of the state were made during the Clarence King's Fortieth Parallel Survey which traversed Nevada during 1867, and during George M. Wheeler's survey of 1869.

Table 4.2. State of Nevada Land Managers and Their Domains

Agency	Acronym	Type of Land	Examples
Nevada Division of Parks	NDP	Nevada State Parks	Lake Tahoe, Valley of Fire, Cathedral Gorge, Beaver Dam
Nevada Department of Transportation	NDOT	Highways and their easements	I-15, I-80, US 95, US 93, SR 167, etc.
Nevada Division of Forestry	NDOF	Highway easements, conservation and public safety in state parks, county parks, and federal land	No land directly administered by this agency
Nevada Division of Wildlife	NDOW	Wildlife Management Areas (WMA)	Overton WMA
University and Community College System of Nevada	UCCSN	University grounds and research areas	Little Valley Research Area (UNR), Rainbow Gardens Geologic area (UNLV)
Nevada Counties		County Parks	Spring Mountain Ranch, Sunset Park, Galena Creek, Topaz Lake

In the late 1930s, biologists from UC Berkeley and the USDA Forest Service led the first vegetation mapping expeditions in Nevada, but these were restricted to the far western portion of the state, along the boundary with California. It was not until the middle of the 20th century that the first general vegetation maps were made for specific areas within the state by Nevada Medal winner W. Dwight Billings. The first general map of Nevada's vegetation was made in the 1970s by the Nevada Department of Conservation. Along with these regional-scale mapping efforts, specialists were constructing distribution maps for the organisms in which they were most interested.

In 1992, the Biological Resources Research Center (BRRC) was established as part of the Department of Biology at the University of Nevada, Reno, to conduct scientific research and planning efforts necessary to preserve the distinct biotic diversity of Nevada. For the past 13 years, scientists at BRRC have been documenting and mapping species in the state, inputting the data into a Geographic Information System (GIS) and allowing the biological data to be displayed with other data such as land ownership, land use, topography, and other geographical data.

The BRRC is a member of the Nevada Biodiversity Research and Conservation Initiative, a collaborative effort among local, state, and federal agencies to incorporate biotic diversity conservation in public land management. The Nevada Biodiversity Initiative is a research project whose goal is to assemble, analyze, and interpret data on biological resources in Nevada and to provide this information in a usable format to the agencies that have the responsibility to manage these resources. The initiative was developed to help integrate long-term planning for biological resources with public land management. The BRRC and the Center for Conservation Biology at Stanford University work to develop and refine baseline biological data. Partner resource management agencies including the Nevada Division of Wildlife, U.S. Bureau of Land Management, U.S. Fish and Wildlife Service, U.S. Forest Service, and the Nevada Natural Heritage Program use the data to develop land management practices. The intention is that these practices will allow for the conservation of biodiversity, maintenance of ecological services, and continued economic viability.

By the end of the 20th century, the first truly modern map of Nevada was created as part of a new national program, the Geographic Approach to Planning for Biodiversity, also known as GAP analysis, the Gap Analysis Program, or simply GAP. Whatever the name, the intention of the ambitious project was to produce high-quality vegetation maps of all US states. The map produced for Nevada in 1997 is a remarkable achievement, mapping the entire state in terms of 65 cover classes (types of vegetation), with a minimum mapping unit size of $100m^2$. The data used were from the Landsat Thematic Mapper satellite, and these were classified and placed into a GIS database.

Unfortunately, one of the main lessons from this exercise was our ignorance. As the Nevada GAP map is used by land management agencies, numerous problems with it have emerged. Curiously, one of the problems is that 65 cover classes are not sufficient to adequately describe the enormously complex ecologic mosaic that covers the landscapes of Nevada. Another interesting outcome was that Nevada is so enormous, that it was exceptionally difficult to verify the occurrence of the cover classes in the locations mapped by GAP. As field teams go out to find the mapped communities, the problems continue to be documented. As a response to this problem in Nevada and other arid western states, a ReGAP program has been initiated that will re-map the state using more people on the ground, a new Landsat satellite, faster computers, and refined classification techniques. Regardless of the public or private entity that we think ought to be managing land in Nevada, it will need to have an accurate map of the distribution of biological resources—the ecological mosaic.

Philosophical Mosaic

Given the complex natures of the land management and ecological mosaics, it should come as no surprise to us that there are many different philosophical views concerning Nevada's environment. We can consider five such viewpoints in very general terms here: Wise Use, Conservationists, Preservationists, Environmentalist "Wackos," and Native American.

The Wise Use Movement, constituted mainly by people in the western US, believes that natural resources should be used. The movement can be considered to be a more moderate arm of the Sagebrush Rebellion, launched in 1979 in Nevada with the "shot heard 'round the west" (Colorado Governor Richard Lamm, 1982), and another volley "fired" with the "shot heard round the world" (Nye County Commissioner Dick Carver, 1994) in 1994 at Jefferson Summit, Nye County, Nevada. They rightfully contend that natural resource exploitation led to the settling and development of the western US. They also resent federal control of the vast majority of the state. Adherents are mainly rural residents, living in close association with the land, many of whom are descendents of the original settlers. In Nevada, many farmers, ranchers, and miners and their associated enterprises agree with the tenants of the Wise Use Movement (although few would agree that they were part of any named movement). People in agreement with this point of view see the federal government as exerting too great an influence on the daily management of lands in the state, and that state and local authorities should have much more, if not full and outright, control. Further, many believe that significant portions of public land should be offered for sale by the federal government to the public, to allow for continued economic growth and development. At its extreme expression, we see ranchers who refuse to recognize federal authority over land management and wind up in jail after losing in court and/or paying fines.

Preservationists believe that natural resources must be preserved, and used only in times of great need when no other alternatives exist. To preserve resources means that we do not use them at all. This is distinctly different from the beliefs of conservationists (see below), who think resources should be used, but not used up. Preservationists frequently advocate the benefits to lands formally designated as Wilderness, National Wildlife Refuges, and National and State Parks, and are concerned about the extinction of any native Nevada plant or animal species.

Conservationists believe that natural resources should be used, (again, I am generalizing) but not to the extent of the use proposed by members of the Wise Use Movement. Conservationists attempt to tread what they see as a "middle ground" between that of the Preservationists and Wise Use Movement factions, in that they recognize that natural resources are the source of wealth and a healthy economy, but are also concerned about sustained and continued use of these resources. While most members of the Wise Use Movement contend that they also are concerned about sustained use, nevertheless, they will disagree with conservationists about the scale and degree of use they find acceptable.

We should also recognize what national radio personality Rush Limbaugh calls "environmentalist wackos," because individuals fitting his definition also reside in our state. We could say that environmentalist wackos are extreme Preservationists, willing to make great personal sacrifices to stop development on natural lands at any cost. If there were major logging operations in the state, these people would be driving spikes into marked trees, an exceptionally dangerous sabotage practice that can result in the death of the chainsaw operator. Members of the Earth Liberation Front (ELF) can be rightfully considered environmentalist wackos, as they will conduct arson, detonate bombs, or otherwise destroy private or public property when they have objections to how it is being used. ELF members, in spite of destroying property, pride themselves on destroying property in such a way that no vertebrate or human life is lost.

No discussion on the different philosophic views concerning Nevada's environment would be complete without also considering Native American perspectives. We have already introduced the land ownership controversy surrounding the Western Shoshone, but there are also Paiute (Northern and Southern), Washoe, and Mojave tribes in the state. Unlike the Western Shoshone, these tribes have all settled their land ownership claims with the US government (there are other issues such as water and grazing rights with which they are struggling). Between the Northern Paiute, Southern Paiute, and Western Shoshone are spoken six of the seven Central Numic Languages, of which the 7th is Comanche. The Comanche left the northeastern border of the Great Basin region once the tribe acquired the horse, and went out into the Great Plains to hunt bison. The Washoe and Mojave languages are not Numic, and have origins outside the region. The Washoe language is related to languages of tribes that are native to the western slope of the Sierra Nevada range, in present-day California. Mojave tribal lands occurs in Nevada only in the extreme southeastern tip of the state, south of Laughlin at the California/Nevada/Arizona border. Today, not all Nevada Native Americans live on reservations or colonies. Just how many live on or off reservations is not known, but there were two estimates produced in 1970, one from data collected by the Bureau of Indian Affairs (BIA) and one from data from the US Census in the same year. The total number, 19,563, is the same, but according to the BIA, 70.1% lived on reservations, while according to the US Census, only 37.1% lived on reservations (Joy Leland in d'Azavedo, 1986).

In spite of their different languages, the Nevada environment was the overriding similarity that bound them together to produce a geographic region called the Great Basin Culture Area. No matter how different anthropologists have drawn the fuzzy boundaries on their maps, all of Nevada is included in the Great Basin Culture Area. In spite of their differences, Native Americans in Nevada shared a successful culture capable of sustaining themselves in a challenging environment. In 1873, about 21,500 Native Americans lived in the Great Basin Culture Area, and it is safe to say that at least 15,000 of these were in Nevada. These people represent a remarkable human achievement, as they and their ancestors found a way to extract resources from the land for thousands of years in such a way as to leave it in what we call now "pristine" condition. Their lives were not easy. They lived in a harsh land without modifying the land, but finding all that they needed there. Central to all Great Basin Native Americans was a worldview that fostered great respect for the land and the resources they extracted from it. Ceremonies, dances, and prayers were offered to the plants and animals whose lives they took to sustain their own.

When a Mosaic's Tiles Don't Fit

When different parties with conflicting interests in the state have such different views on the environment and the human role in that environment, then there are bound to be conflicts concerning the use of the land. There is no shortage of contentious environmental issues in the state to examine, but we will consider four such cases here: Bighorn sheep, wild horses and burros, Clark County growth, and the Lake Tahoe Basin.

Bighorn Sheep

Often nature-loving Nevadans disagree about specific management goals of the state's wild lands. For instance, there are various opinions concerning the Nevada State Animal, the Bighorn Sheep (*Ovis canadensis*). In the early 1800s, there were perhaps about 2 million individuals throughout the west, but by 1975, only 10,000 individuals remained. Today, the Desert Bighorn Sheep are making a strong comeback, due largely to efforts by Nevada Bighorns Unlimited in collaboration with the Nevada Division of Wildlife and the US Fish and Wildlife Service. Tree-huggers (sheep-huggers too) take note: this private hunter's group has been paramount in the recovery of this species, and its reintroduction to many locations in which the populations were **extirpated**. We might think that everyone would be happy with these activities, but we would be wrong. In fact, we find resistance from a group of Nevadans who are deeply committed to a healthy environment: sheep ranchers. Why do you suppose some in this group oppose the widespread reintroduction of the native sheep? Can you think of any biological or ecological reasons? Could they have political objections?

Wild Horses and Burros

Large plant-eating mammals like sheep frequently evoke warm, fuzzy feelings from people toward them. In the American West, the horse is particularly evocative for local residents. While the Native Americans in Nevada did not use horses for most of their history, the horse was essential to the European settlement of the state. To Nevada cowboys, a horse is their most important possession. Wild horses and burros occur in all 17 Nevada counties; in fact, 40,000 wild horses live in Nevada, more than half of the nation's total. Wild horses and burros are protected in the US by Public Law 92-195, the Wild Free-Roaming Horse and Burro Act, passed unanimously by both the US Senate and House of Representatives.

The Federal Land Policy and Management Act, dated Oct. 21, 1976, amended the original act, and allowed for the Secretaries of the Interior and Agriculture to use helicopters and motorized vehicles to manage wild horses and burros on public lands. The Public Rangelands Improvement Act of 1978 established that rangeland must be inventoried, identified, and monitored in order to manage and improve all rangeland values. It reaffirmed that wild horses and burros had a right to live free from human harassment, but that excess horses should be removed. It established an adoption program where individuals could gain title to excess horses gathered by the program.

In 2001, a bill was introduced from children in a fourth-grade class Roberts Elementary School in Henderson to make the Wild Mustang (horse) another Nevada State Animal. Upon passage, Assembly Majority Leader Barbara Buckley, who introduced the bill on behalf of the class, called the mustang the "emblem of the America spirit." Even though the bill passed, two representatives voted no, and a contentious debate for many days was covered by the state's newspapers. Why do you suppose there was such a debate? This was the legislature, so were the reasons for debate strictly political? Or, were there biological or ecological reasons for the disagreements? Can you think of any biological or ecological reason why wild horses and burros should not be on Nevada's rangelands?

There is a reason to keep the horses off the range from the perspective of an **ecocentric** (not an eccentric, even though you may consider such a person to be so). However, from the **biocentric** worldview, there is a reason to have them on the range. If the viewpoint is **anthrocentric**, I can think of at least one reason why they should be there, and at least one reason why they should not. An excellent exercise for you to strengthen your knowledge of other worldviews and your **critical thinking** skills is to consider this issue. Find all plausible reasons any viewpoint could propose for having or not having wild horses on public lands. After writing down the reasons you found from any worldview, do some internet surfing or library work to find the reasons that advocates on both sides of the issue are citing. Next, analyze the reasons and pit them against each other in your mind. Weigh the evidence and arrive at a tentative decision. Congratulations! You have used critical thinking to make a decision regarding an issue concerning the management of Nevada public lands. It is a tentative conclusion because you may change your mind later, depending on additional evidence gained or arguments offered. That's critical thinking; check it out, it's cool.

Clark County Growth and the Multi-Species Habitat Conservation Plan

Although the amount of land that the US Fish and Wildlife Service (USFWS) manages in NWRs is relatively small (Table 4.1), this agency is responsible for the preservation of species wherever they occur in the state, with authority granted to it by Congress under the amended Endangered Species Act of 1973 (ESA). To preserve species, it is necessary to preserve their **habitat**, or where they live. In the Las Vegas Valley, there is concern that 79 species may be vulnerable to extinction given the explosive growth in the valley (See Chapter 2) that consumes the habitats of all these species in its wake (for us to live here, someone has to go!). In order for the agency to allow the habitat destruction of these vulnerable species, the USFWS conditionally granted permission to Clark County to do so because the county assured the agency under their Multi-Species Habitat Conservation Plan (MSHCP), any habitat destroyed would be replaced elsewhere in the county, thus conserving the species. The approved plan is one of the most progressive plans, if not the most progressive plan of its kind in the world. It involves a consortium of local, state, federal agencies, public interest groups, and the state's institutions of higher education. Subject to constant monitoring and debate, the MSHCP is a work in progress that is followed closely by people interested in conservation throughout the world.

Lake Tahoe Basin

Lake Tahoe is an enormous lake of striking beauty that straddles part of the central California/Nevada border. At 6,200 feet elevation in the high Sierra Nevada mountains, this lake is 23 miles long, 13 miles wide, and 1645 feet deep. The Truckee River originates as overflow from the lake, supporting the communities of Reno and Sparks in the Truckee Meadows. Farmers along the river and in Fallon also depend on the Truckee River, as does Pyramid Lake, where the Truckee River terminates. From Lake Tahoe through the Truckee River's course to Pyramid Lake, there are many species with special status under the ESA. Tourism is the main industry at the human communities at the lake; a tourism dependent upon the natural beauty and proper functioning of its landscape's ecosystems. On this stage, an environmental bombshell was dropped: the lake's world famous clarity was declining at an alarming rate.

In 1969, Nevada and California formed a bi-state compact, approved by the US Congress, the

Tahoe Regional Planning Authority (TRPA). Local governments and non-profit organizations are included in the compact, and work closely with federal agencies. One of the central tenants of the compact is that human use of the environment needed to be balanced and regulated. The TRPA was far ahead of its time in terms of the complexity of the jurisdictional mosaic and the determination to cooperate in order to preserve the unique qualities of the area. The compact is more difficult than the Clark County MSHCP in that there are two states and five counties involved. As a result, the TRPA has been mired in difficulties since its conception.

Many issues are contested in the basin, and several involve how people travel in the area. By definition, the tourist industry needs tourists. Most use automobiles, adversely affecting the air quality. Thus, the visitation deteriorates the aesthetic quality of the lake that brought the tourists there in the first place. While there is a regional bus system in place, most visitors and locals prefer to have their own vehicle. Traffic on the lake and in the mountains are also contentious issues. Should people be allowed to use personal watercraft and snowmobiles if they adversely affect water quality and wildlife? Mountain bikers, equestrians, hikers, and four-wheelers agree that the forests should be beautiful and have well-maintained trails. Unfortunately, most would rather not see each other, regardless of the means of conveyance. Occasionally, people traveling by one method are belligerent to those traveling by another method. With increased recreational demands, it is becoming more difficult to find places to travel where you see no one else.

All parties want "forest health," but agreement is elusive as to what forest health means. For instance, to loggers, a healthy forest is one in which we can harvest timber to supply building and fuel needs. To protect the resource and keep it healthy, loggers think dead trees should be cut, fires should be prevented, and sustainable logging operations should be allowed. To hunters, a healthy forest is good deer habitat. Periodic, low level fires improve deer habitat by creating openings in the forest and promoting plants that deer like to eat. To birdwatchers, a healthy forest has young and old trees along with dead ones, the dead trees are resources to be used by cavity nesting birds. To ranchers, a healthy forest is one in which fence and corral posts are available. The forest's vegetation produces feed for cattle, holds the soil in place, and slowly captures water, in order to deliver it to the lowlands in the summer. To still others, a healthy forest is highly diverse, with many forms of life. To others, the composition of the forest is not nearly as important as its beauty. To these people, healthy forest is a beautiful forest, in whatever terms that they define beauty. These different viewpoints are not necessarily compatible, yet they live together in the same forest.

Sometimes, people are locked in direct conflict over these issues in the courts, as was recently brought to national attention in the Supreme Court. In this case, a temporary 32 month moratorium imposed by the TRPA in 1981 prevented hundreds of landowners from building homes. The lands are environmentally sensitive, and since the lake's water quality and other indicators were not improving, a court order extended the moratorium for many years. Finally, a regulation prevented about 400 landowners from building their homes. These landowners contended this was a "taking" of private property. Further, since they were not compensated for the taking, this taking was a violation of the Fifth Amendment. The court decided 6-3 in favor of the TRPA on 23 April 2002. The majority opinion was that the needs of a regional planning agency to make decisions affecting communities outweighed the right of an individual to control a parcel in the community.

However, in spite of the TRPA partnership's efforts, by 1997 the water clarity had declined even further. This moved President Clinton to pronounce the lake a "national treasure," jumpstarting an intensification of efforts in both basic research and restoration efforts from salvage logging and prescription fires to runoff control. It remains to be seen what balance will be found between human development and the Lake Tahoe ecosystems.

CHAPTER 5

Nevada Agriculture

Amina Sadik and David A. Charlet

Historic Importance

The importance of agriculture in Nevada is not obvious, as visitors flock mainly to the neon cities, where showgirls are decked out in stars and feathers. Nevertheless, agriculture has always been one of the major industries in Nevada, with its contributions to the Nevada economy today ranking behind only tourism and mining. During the initial European settlement of Nevada with the building of missions by the Church of Latter Day Saints in 1850, agriculture was the primary enterprise. Settlers had to grow their own food because of Nevada's isolation from California to the west by the high passes of the Sierra Nevada and because of the vast distances to agricultural centers east of the Rocky Mountains. Because the hardy settlers could grow food, the settlements thrived.

After the discovery of what were the world's richest deposits of gold and silver at the Comstock Lode near Virginia City in 1859, the mining operations were successful because food was grown for the miners in the nearby Carson and Eagle Valleys and in the Truckee Meadows. Agriculture was the base that supported the mining, allowing Virginia City to become the largest and richest city west of the Mississippi River. Throughout the 20th century, virtually all land in Nevada was used to support agricultural activities (Figure 5.1, Figure 5.2).

Nevada's agricultural industry can be divided into two types: ranching and farming. Ranching involves the management of livestock, whereas farming uses land to grow crops. Farming takes place in Nevada where irrigation is possible through either stream diversion or underground pumping.

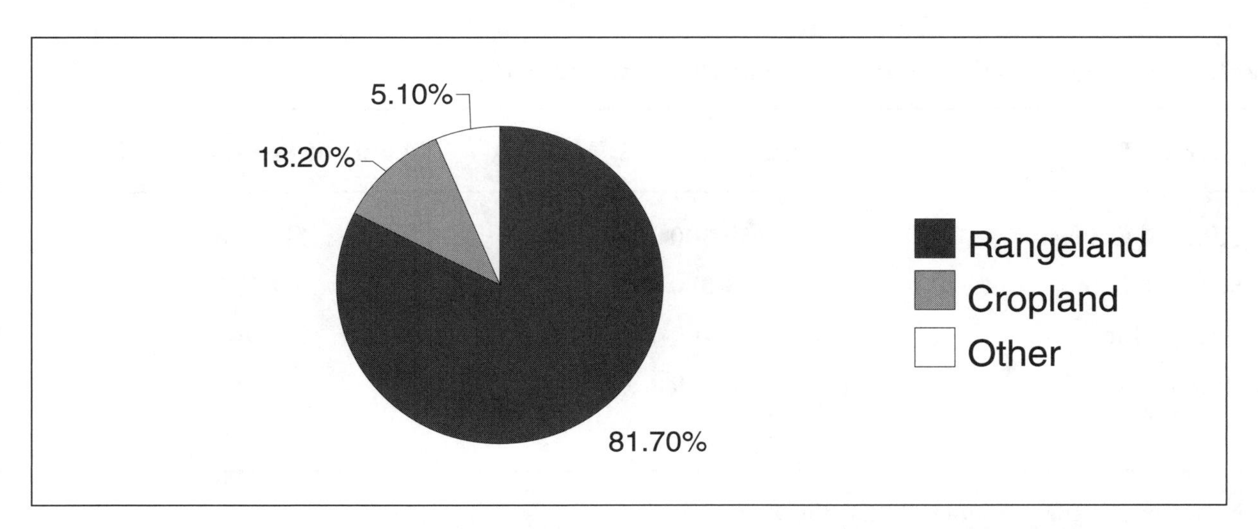

Figure 5.1. Land Use in Nevada. Note that rangeland total was calculated before certain public lands have been taken out of service.

Ranching

Small in number, Nevada ranches are ranked third in size in the nation. According to the Nevada Agricultural Statistics Service (NASS), they average 3,500 acres. The state's agricultural industry is focused on ranching because of the availability of grazing areas provided by mountains surrounding the valleys. The raising of livestock represents the most important part of Nevada's agricultural industry (Table 5.1). Nevada livestock includes beef and milk cows, calves, sheep and lambs, milk and Angora goats, hogs and pigs, chickens (layers and pullets), turkeys, minks, honeybees, horses, emus, and llamas. As one might expect, due to animal feed availability (pastures, alfalfa production), the northern part of the state raises the highest concentration of cattle. Elko County occupies second place in the nation among counties producing beef. Dairying is a growing industry in the state. This industry is most abundant in the valleys near Las Vegas and Reno. More than half of the ranches in Nevada raise sheep and lambs; however, production is concentrated in the northern part of the state. The production of hogs and pigs is limited to the local market. Mink are raised for their pelts. Honeybee hives are often placed in alfalfa fields for the production of honey. Horses have always been important in Nevada, but recently have become a lucrative business in the state. They are used for both work and pleasure.

Farming

By far the most important crop produced by Nevada farms is hay, accounting for nearly 20% of all agricultural receipts in the state. Other major crops produced in Nevada include alfalfa seed, root crops such as onions, garlic, and potatoes, and grains such as corn, wheat, barley, and oats. The arid, high desert climate of Nevada offers perfect conditions for high quality crops when water is available to them. Garlic and onions do well in Nevada, as they grow during cool months, reducing their water demand. Potatoes require significant irrigation, but the quality of Nevada potatoes is excellent.

The single most important plant in Nevada agriculture is alfalfa (*Medicago sativa*), which is grown not only for its hay (alfalfa hay is the leading cash crop in the state), but also for its seed. Even though alfalfa is a native of the Nile River in Egypt, it does remarkably well in Nevada when it is given about three acre-feet of water per acre each growing season. An acre-foot is 326,000 gallons, or a pool of water one foot deep spread over each acre of the field (equivalent to 36 inches of annual precipitation, or about five times the annual precipitation in Reno). Needless to say, alfalfa hay is produced in the parts of the state where irrigation is possible throughout the long growing season. Nevada farms produce some of the finest hay in the world, with cuttings taken

Table 5.1. Major Agricultural Products and Their Values for 1999[1]

Year 1999	**Production** *(000's lbs)*	**Value of Production** *(000$'s)*
Cattle & Calves	159,500	111,308
Sheep & Lambs	3,080	2,022
Hogs & Pigs	2,709	945
Honey	9	383
Milk	497,000[2]	65,436[3]

[1]Courtesy of Nevada Agricultural Statistics Service.
[2]Production includes 5 million pounds used on farms; the remaining is marketed.
[3]Value of production indicates cash receipts from marketing.

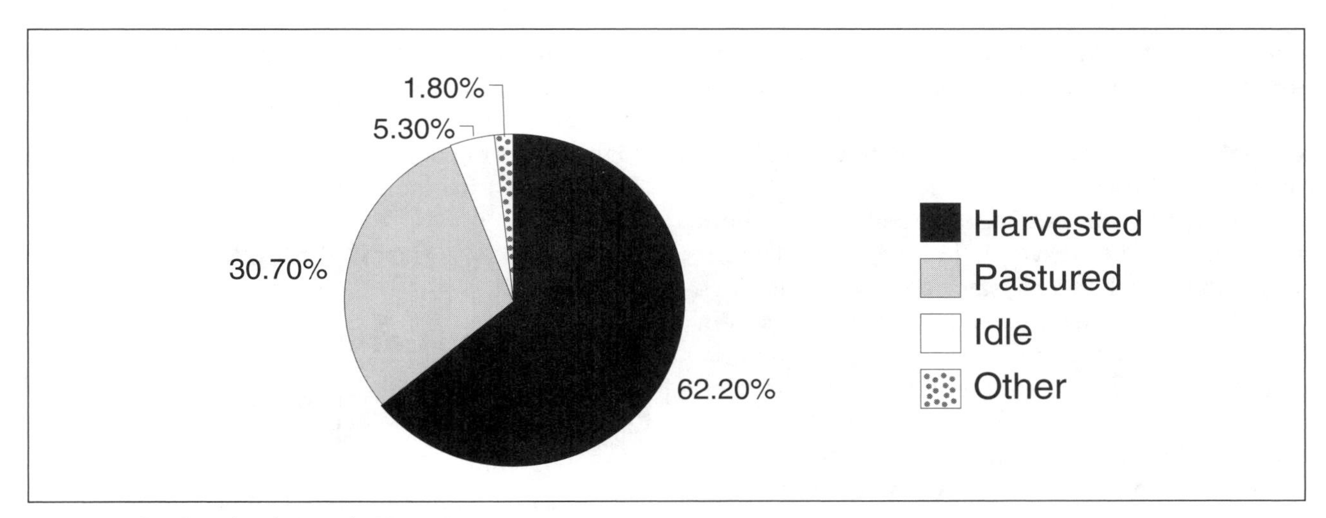

Figure 5.2. Cropland Uses in Nevada.

during the cooler parts of the growing season yielding very high total digestible nutrient (TDN) values and high protein. These qualities make for premium dairy hay.

Several cuttings of alfalfa hay are made during the year. Because of the longer growing season in southern compared to northern Nevada, southern alfalfa farms can harvest up to five crops per year. Southern Nevada alfalfa operations can begin their first cutting in May and cuttings can continue into late October. The best hay comes from northern Nevada, where farmers often get only three crops per year, cutting in June, late July, and September. The first and third crops, when cut during the early stages of bloom, provide the best revenue as they are premium-grade dairy hays. Holstein dairy cattle fed with northern Nevada first crop hay and grain commonly provide more than seven gallons of milk a day. Second crop hay is used to feed dry stock such as bulls and calves. While pure alfalfa hay is best for dairy cattle, with California dairies accounting for the lion's share of the domestic market (Figure 5.3), hay made from pasture grasses mixed with alfalfa and/or clover is better for horses. Nevada farmers grow these hays as well.

The plants used in hay crops are all perennials. As such, they do not need to be planted every year. Even though alfalfa hay production peaks on a field within a decade, if circumstances (such as drought) prevent replanting, fields can still be maintained and harvested at lower levels for 20 years or more. Hay is harvested in several steps. First the fields are cut with windrowers that move a sickle forward through the field, cutting the stems several inches above the ground, and leaving the cut hay in a narrow pile, or windrow, behind the machine. After the hay dries in the field,

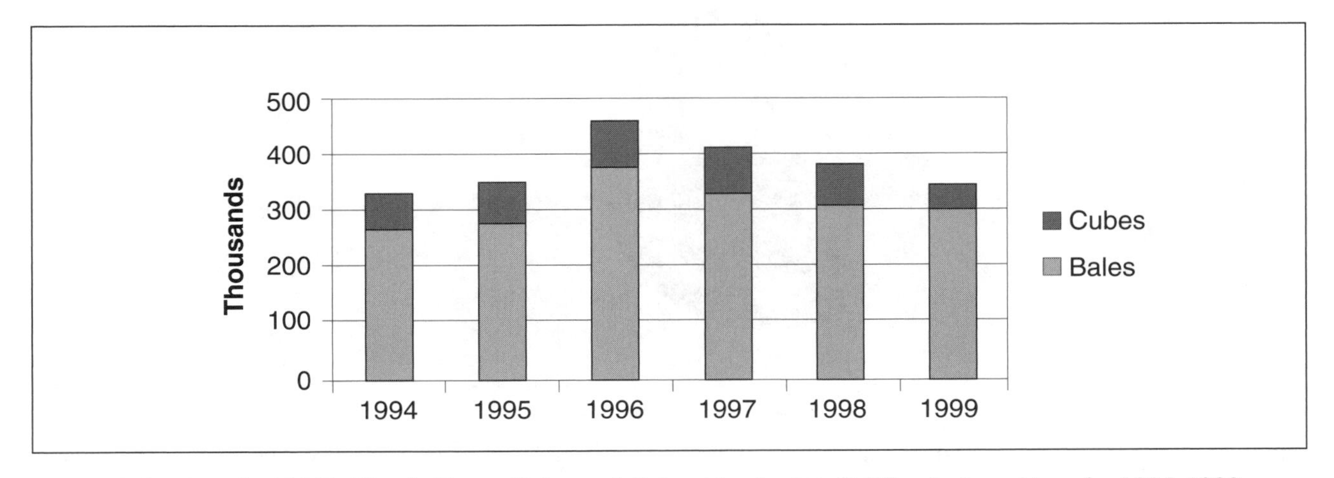

Figure 5.3. Nevada Alfalfa Hay in Tons. Bales and Cubes Trucked to California from Nevada, 1994-1999.

a baler is sent down each windrow. The baler picks up the hay, chops it, compresses it, and wraps it with twine or wire into a "bale," spitting the bales out behind it. Another specialized tractor called a haro-bed follows the baler, picking up the bales and stacking them onto a large bed that can hold up to 64 bales that weigh 110 pounds. Then it drives them back to a haystack and stacks the bales neatly into the expanding stack. If the stack is in the open, an extra set of bales, usually of oat straw (the stems of oat plants without the seed), is placed over the top. When it rains, most of the rain is absorbed by these oat bales, preventing the good alfalfa hay from spoiling. Hay stored like this can remain good for a year or more in Nevada. Recently, many Nevada farms have stopped putting their hay in bales, but instead into giant bales called "hay cubes," weighing about several hundred pounds each. Poor quality hay, such as hay that has lots of weeds like foxtail barley that have much protein but whose stickers can infect livestock mouths, is ground into pellets and exported overseas, mainly to Japan (Figure 5.4).

Nevada farmers also produce alfalfa seed; however, in order to do so they must allow the plants to mature beyond the point where they make premium hay. In addition, to ensure good seed crops, carpenter bees are introduced to serve as pollinators. Large artificial nesting areas are brought to the fields to support the large populations of bees necessary to insure complete pollination, and thus the "setting," or development and ripening, of the seed. So, if you are an alfalfa farmer in Nevada, you either try to grow your best dairy hay or seed. Large alfalfa seed operations occur near Lovelock, Nevada.

Root Crops

Onions and Garlic

Onions and garlic are produced in northwestern Nevada (Lyon, Washoe, and Humboldt counties) for both fresh market and dehydration. Onions are planted in March and harvested as onions and onion seed in September. While the production of onions has increased in Nevada since 1993 by nearly 300%, the dollar value of the yield has remained about the same (Table 5.2). Although garlic is mostly grown for seed, some garlic is also dehydrated. Garlic harvest begins in July and continues into the fall. From 1993-1999, Nevada farmers have managed a threefold increase in garlic seed production (Table 5.3), and the total value of their crops has more than doubled.

Potatoes

Seven thousand acres of potatoes are planted in April with the first precipitation. According to the USDA, the average yield per acre in 2000 was 381 cwt., for a total yield of 3.15 million cwt. state-

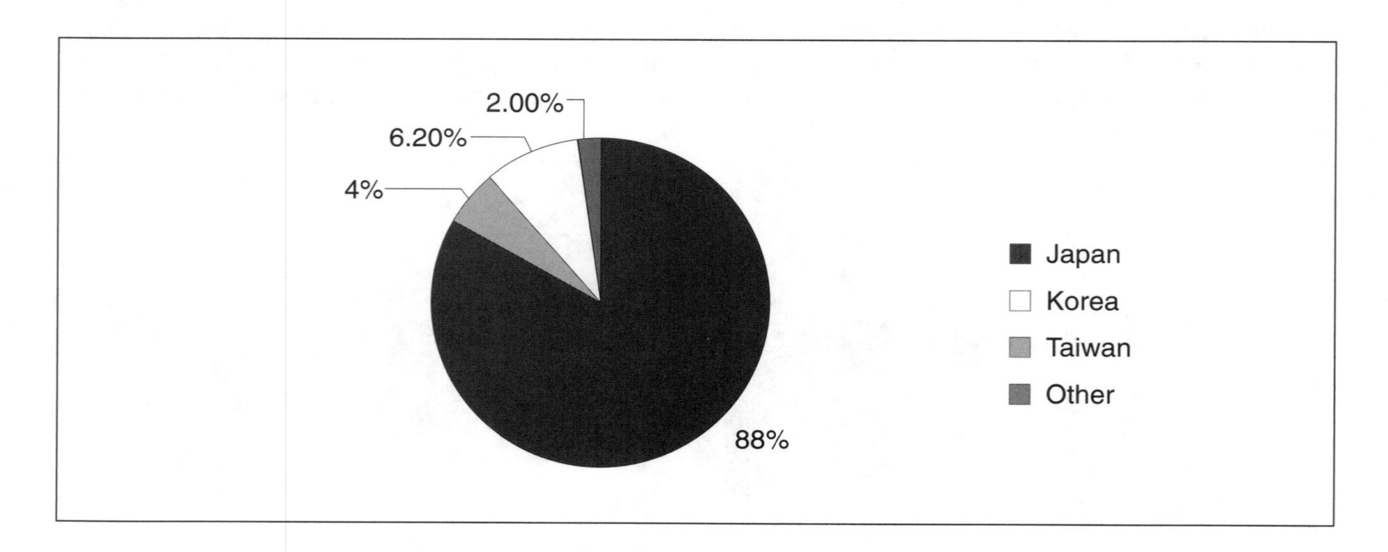

Figure 5.4. Nevada Agricultural Exports for 2000, by Weight. Alfalfa hay accounted for 93% of the total poundage exported.

Table 5.2. Nevada Onion Production[1]

Year	Acres Planted	Production *(Tons)*	Value of Production *(000$'s)*
1993	1,000	28,000	15,120
1994	1,500	31,500	8,820
1995	1,900	43,700	11,362
1996	1,900	55,100	15,979
1997	1,800	45,900	13,311
1998	2,100	46,200	12,938
1999	3,200	78,400	13,328

[1]Courtesy of Nevada Agricultural Statistics Service.

Table 5.3. Nevada Garlic Production for Seed[1]

Year	Acres Planted	Production *(Tons)*	Value of Production *(000$'s)*
1993	800	6,400	1,984
1994	1,650	12,375	3,651
1995	1,650	12,705	4,447
1996	1,800	14,220	5,261
1997	2,100	15,750	5,198
1998	2,300	16,560	5,134
1999	2,600	16,900	5,070

[1]Courtesy of Nevada Agricultural Statistics Service.

wide. Total potato production in that year was estimated at $2.59 billion (Table 5.4). Potatoes generally start germinating in May under cool conditions and blooming in July. Digging starts in August and continues into the fall.

Specialty Crops

Specialty crops include such products as mint, fruits, and nuts. Fruits such as apples, apricots, cherries (sweet and tart), grapes, pears, plums, and prunes are grown in northern Nevada. In southern Nevada, grapes, pomegranates, pecans, and pistachios are grown. One farm in Pahrump Valley in Nye County produces wine from the grapes it grows.

Nevada Soils

The Orovada soil is the most widespread soil in Nevada, and that most often used to grow alfalfa, potatoes, corn, wheat, oats, and barley. It is found on more than 360,000 acres in northern and central Nevada. It is a rich, deep soil usually found in the wild in areas dominated by sagebrush. Farmers in northern Nevada use big sagebrush as an indicator of good soils to use for agriculture.

Can Agriculture in Nevada Be Sustainable?

Nevada presents a challenging climate for agriculture. For one thing, it is the driest state in the country, and precipitation can vary wildly from year to year and from place to place. Nevertheless, the hardworking and ingenious Nevada farmers have applied clever, low-tech solutions to many problems they have faced in 150 years of Nevada agriculture. Nonetheless, problems remain, and much of the best farming land has been taken out of production in order to make room for our fast-growing cities. For our arid climate to support sustainable agriculture into the future, at least three options are being developed that can increase production and minimize the environmental effects of farming: hydroponic greenhouses, drought-resistant farming techniques, and the development of salt-resistant crop plant strains.

Hydroponics

Hydroponics, from the Greek words *hydro* (water) and *ponos* (labor), is used to mean "growing plants in water." Hydroponics is a method of growing plants with their roots in a solution containing a balanced proportion of nutrient elements essential for plant growth. Plants need soil only as a support and as a source of inorganic elements, water, and air. Thus, with or without that support, any source fulfilling these needs would allow plants to grow and develop. Hydroponics offers a solution in areas where the soil is poor, thin, compacted, and impermeable.

There are two types of hydroponics: with and without support. In the first case, a type of inert, artificial soil is used that may include a mixture of sand, vermiculite, or perlite with organic compounds. This support is supplied with water containing all the inorganic nutrients the plant needs at any particular moment during the course of its development. Usually, the plant's support mixture is contained in a plastic tubular grow-bag that prevents water evaporation. Each plant takes root through one of many perforations (approx. 5 cm. in diameter) in the tube. The plants are vertically maintained by strings from above (Figure 5.8) that are manually and periodically drawn tight as the plant grows. In the second case, no rooting medium is used for support. The plant's root system bathes in a film of nutrient-rich water contained in a long plastic tube, which allows the water to circulate back to the nutrient-rich water supply source.

One of the agricultural producers using hydroponics with support is a tomato grower located in North Las Vegas. Although we use tomato production as an example, herbs, cucumbers, lettuce, and other vegetables are also produced in Nevada

Table 5.4. Nevada Potato Production 1993-2000[1]

Year	Acres Planted	Yield per Acre *(Cwt)*	Production *(000 Cwt)*	Value of Production *(000$'s)*
1993	7,700	380	2,926	16,239
1994	8,000	345	2,760	16,974
1995	7,600	365	2,774	23,024
1996	8,000	400	3,160	10,902
1997	7,000	430	2,967	13,352
1998	7,000	400	2,800	11,900
1999	6,500	440	2,860	14,014
2000	7,000	450	3,150	16,002

[1]Courtesy of Nevada Agricultural Statistics Service.

greenhouses (Klein 2001). In the tomato operation, custom-made grow-bags containing perlite (a recyclable mixture of volcanic rock) are used as a support, in which four plants are rooted. At the level of the root system, capillary tubing delivers the nutrient-rich solution (nutrient drip) needed by the tomato plant for optimal growth (Figure 5.6). Years of accumulated knowledge and experimentation have yielded special nutrient recipes using different water-soluble fertilizers. The optimal concentration of the different elements included in a nutrient-rich solution (Table 5.6) concocted by Ken Gerhart of Sunco Ltd., contains different proportions of the elements included in the different chemical origins listed below.

Composition of Viking Ship Calcium Nitrate Hydro Agri Solution Grade:

- Ammoniacal Nitrogen 1.0%
- Nitrate Nitrogen 14.5%
- Total Calcium 19.0%

Champion Greenhouse Grade:

- Nitrate Nitrogen 13.75%
- Soluble Potassium 46.0%

Giles:

- Total Magnesium 9.8%
- Water Soluble Magnesium 9.8%
- Combined Sulfur 12.9%

Vicksburg Chemical Company:

- Available Phosphate (P_2O_5)............ 52.00%
- Soluble Potash (K_2O) 34.00%

Table 5.5. General Agricultural Production in Nevada in 1999[1]

Year 1999	Acres Harvested *(000's)*	Yield per Acre *(Units)*	Production *(000's)*	Value of Production *(000$'s)*
Alfalfa Hay	255	4.1 tons	1,046 tons	87,341
Alfalfa Seed	13	690 lbs	8,970 lbs	12,109
All Other Hay	225	1.8 tons	405 tons	31,590
All Wheat	15	91.7 bu	1,375 bu	3,639
Barley	4	90 bu	360 bu	846
Potatoes	6.5	440 cwt	2,860 cwt	14,014
Garlic	2.6	6.5 tons	16.9 tons	5,070
Onions	2.8	28 tons	78.4 tons	13,328

[1]Courtesy of Nevada Agricultural Statistics Service.
bu bushel
cwt hundredweight (100 lbs)

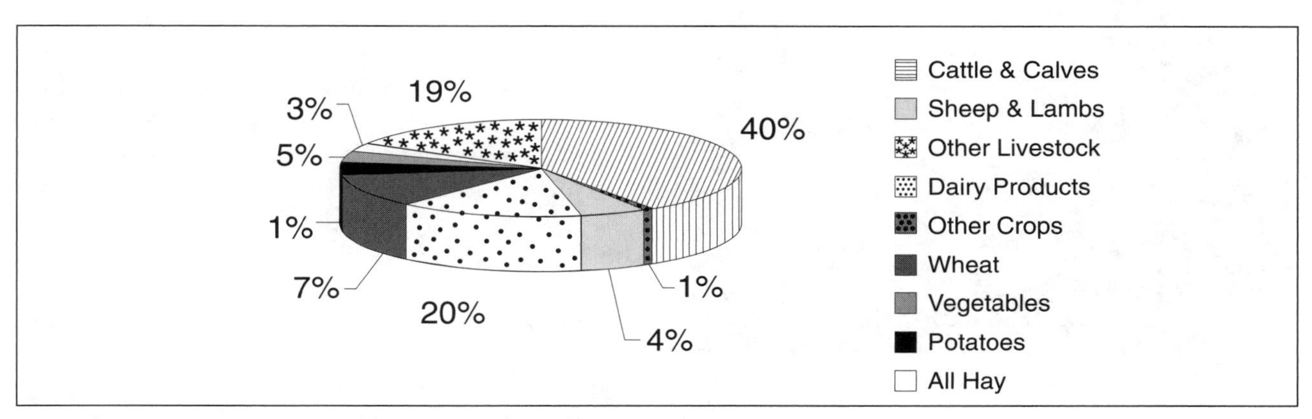

Figure 5.5. Global Cash Receipts of Nevada Farms in 1999.

A successful hydroponics operation needs to provide more than just nutrient-rich solutions to its plants in order to ensure optimal growth and maximum yield. For example, the twelve-acre greenhouse producing tomatoes in North Las Vegas operates under a set of strict environmental controls. Carbon dioxide concentration is maintained between 450 and 500 ppm, and is delivered by way of a perforated hose. It is released as needed from tanks filled with carbon dioxide gas located outside the greenhouse. The relative humidity maintained in the facility falls between 65% and 75%. The average temperature is 70° F. in the daytime, and 65° F. at night. This temperature is maintained at the desired level with the help of a heating system. Because of the challenging weather in Las Vegas, and in Nevada as a whole, the temperature is adjusted using two cooling systems. Roof fences (Figure 5.7) open or close in response to wind direction, and are tinted or clear in accordance with the season. Cooling is also accomplished by way of a high-pressure fog system (Figure 5.7). These various physical parameters are regulated by computer.

Because of the large surface area covered by a commercial greenhouse, the production area must be subdivided into a number of compartments, each of which is monitored by different computer sensors. For example, hygrometry is monitored so that humidity in a given compartment will not exceed a percentage known to favor the development of certain pathogens. As in any other method of food production, hydroponics demands close monitoring of the nutrient source to avoid contamination and resultant pathogenic diseases. However, because this culture is contained in a greenhouse, the number of pathogens is limited compared to open field agriculture. In facilities such as the one in North Las Vegas, it is essential to avoid diseases because the tomato growers rely more on biological controls than on the use of pesticides. Soft pesticides are used only as a last resort because harsh pesticides would adversely affect the hives of the bumblebees (*Bombus occidentalis*) they are using as pollinators (Figure 5.8). As an alternative to pesticide use, these professional growers are using the two principal pests of biological control. For the whitefly, they use Ercal Week wasps, and for the leaf miner (a small insect which, in the larval stage, burrows in and eats the parenchyma of leaves), they use another species of wasp, *Diglyphus isaea*.

Figure 5.6. Grow-bag and nutrient drip shown by Ken Gerhart, President and Co-Manager of the 12-acre Black Hills Energy greenhouse known as Sunco Ltd.

Figure 5.7. High-pressure fog system, adjustable strings, and roof fences (Sunco Tomato Greenhouse).

Table 5.6. Target Values of Nutrient Solutions for Greenhouse Tomatoes

Nutrients	Target Solution (ppm)	Media Optimum[1] (ppm)	Media Range[2] (ppm)
NO_3-N	210	224	170 – 270
NH_4-N	16	< 7	1 – 7
P	47	31	22 – 62
K	293	310	250 – 390
Ca	230	400	240 – 480
Mg	66	110	65 – 155
S	120	220	130 – 290
Cl	53	315	210 – 420
Na	< 25	< 180	20 – 180
Fe	0.9	1.4	0.5 – 2.0
Mn	0.55	0.3	0.17 – 0.55
Zn	0.33	0.45	0.3 – 0.65
B	0.36	0.55	0.4 – 0.7
Cu	0.05	0.04	0.03 – 0.1
Mo	0.05	0.05	0.03 – 0.08
EC[3] (mS)[4]	2.5	3.7	2.8 – 4.2

[1]Optimum Medium is the ideal nutrient mix in the grow-bag.
[2]Medium Range is the area within which the grower wishes to stay.
[3]EC indicates Electrical Conductivity, which is dependent upon salt concentration.
[4] (mS) refers to milliSiemens (mho), which is the reciprocal of the ohm (unit of electrical resistance).

Figure 5.8. Bumblebee hive used for pollination of tomatoes in Sunco Greenhouse.

Knowing that traditional agricultural practices have been shown to be the largest nonpoint source of water pollution, using the technology of hydroponics — where water is contained at all times — is a positive agricultural alternative when possible. Because of the aridity of the climate and the resultant water shortage, using hydroponics technology avoids the problem of pesticides, as well as nitrates from fertilizers and manure, in the ground water. In addition, problems common to arid agriculture, such as salinization and erosion by irrigation, are eliminated. Another positive aspect of hydroponics involves recycling of the artificial soil. For example, after eleven months of use (from planting to final harvest), the perlite is recycled and, together with the removed tomato vines, is used in composting. The resulting compost is sold as potting soil.

Hydroponics, with the recent development of the nutrient film technique (NFT), has been successful in California, Arizona, England, Morocco, Abu Dhabi, Saudi Arabia, and Iran. It continues to allow Morocco to have a sustainable agriculture and be competitive in exporting fruits and vegetables to Common Market countries. It also permits the export of carnations and roses to the United States.

Drought-Resistant Farming

Drought-resistant farming is successful only if drought-tolerant crops are used in drought-prone soils, which is the case nearly everywhere in Nevada. However, soil composition can be modified to capture much of the rainfall in order to store water for future plant use. Such modification includes increasing the organic matter content of the soil, which improves both water retention and aggregation. Aggregation is known to increase water infiltration into the soil (Hudson 1994, Boyle et al. 1989).

Genetic modifications in plants can result in hardier crops. For example, U.S. scientists have identified a gene that controls water retention in plants, thereby increasing drought resistance. The gene responds to a natural acid that controls the size of openings in leaves ("Drought Breakthrough," http://www.thescotsman.co.uk).

Studies in drought-resistant crops are performed not only in the United States, but also elsewhere around the globe. In India, for example, geneticists from the International Crops Research Institute for the Semi-Arid Tropics (ICRISAT) have designed a new drought-resistant variety of pearl millet. ICRISAT staff has been able to speed up the adoption process by working with local farmers to introduce the crop, rather than working only on experimental plots. Radio and television have helped spread the word, and the new variety is being disseminated among farmers faster than ever before (Food and Agriculture Organization, http://www.jhuccp.org/popline/popbib/m13.stm#99).

At the local level, in Las Vegas, the University of Nevada Cooperative Extension maintains an organically grown fruit orchard comprised of over 450 trees and vines, including 11 cultivars of apples, seven of pears, including a bright red cultivar, seven of apricots, ten of nectarines, 13 of peaches, six of grapes, five of cherries, persimmons, pomegranates, figs, and a Chinese date. This program's original objective was to test which varieties of fruit would grow in a desert environment, but the orchard is also used for classes in pruning, training in bare root planting, testing different cultivars with lower chilling hours for desert growing, etc. This year, it is expected that the orchard will produce over 2,400 pounds of fruit. The orchard has a weather station to record chilling hours, wind, and other conditions that affect fruit growing. The data obtained by the station will be used in the future as a reference for valley fruit growers (Klein 2001).

Salt-Resistant Crop Plants

A common problem throughout the world when farming in arid climates is salinization, or the accumulation of salt in soils, and Nevada is not immune to this problem. However, many species of native plants in Nevada have evolved the means to tolerate or even thrive in saline soils. Plant scientists at the University of Nevada and elsewhere are working on means whereby the genes governing these characteristics can be identified, isolated, and transferred to crop plants in order to confer upon them salt-tolerance. The development of such strains could help farmers to use land that otherwise would have to be taken out of production because of the salt content in the soil.

References

Anderson, T.L. and B. Yandle. 2001. Agriculture and the Environment—Searching for Greener Pastures. Hoover Institution Press. Stanford, California.

Boyle, M., W.T. Frankenberger, Jr., and L.H. Stolzy. 1989. The influence of organic matter on soil aggregation and water infiltration. *Journal of Production Agriculture* 2: 209-299.

Hudson, B.E. 1994. Soil organic matter and available water capacity. *Journal of Soil and Water Conservation* 49(2): 189-194.

National Agriculture Statistics Service Internet Homepage. 2001. http://www.nass.usda.gov/nv/

Klein, Margie Bennis. 2001. Program Coordinator, University of Nevada Cooperative Extension (personal communication).

Owens, Martin J. 2001. Nevada Agricultural Statistics Service, statistician (personal communication).

Tan, K.H. 1994. Environmental Soil Science. Marcel Dekker, Inc. New York.

Wolf, B. 1999. The Fertile Triangle. Food Products Press. New York.

United Nations, Food and Agriculture Organization (FAO). 1996. *Food for all.* Rome. 64 pp. http://www.jhuccp.org/popline/popbib/m13.stm#99

Internet Resources

National Agriculture Statistics Service, Nevada. http://www.nass.usda.gov/nv

National Oceanic and Aviation Administration, Las Vegas. http://www.wrh.noaa.gov/lasvegas/

State of Nevada, Division of Agriculture. http://www.agri.state.nv.us

National Resource Conservation Service, Nevada. http://www.nv.nrcs,usda.gov/pr/nevada

University of Nevada, Reno, College of Agriculture. http://www.ag.unr.edu/coa

Contacts

College of Agriculture, Biotechnology and Natural Resources and Nevada Agricultural Cooperative Extension Service 702-222-3130

National Weather Bureau 702-263-9744

Definitions

bu: bushel

cwt: hundredweight; 381 cwt = 38,100 lbs.

idle: land that is either not being cultivated, or is unsuitable for cultivation.

strass: a brilliant glass of high lead content used for the manufacture of artificial gems.

CHAPTER 6

Environmental Health

Marc L. Rigas

Supplemental Article

Shane Synder

DISCLAIMER: *The author is a scientist with the Office of Research and Development of the U.S. Environmental Protection Agency. This paper has been reviewed in accordance with the Agency's peer and administrative review policies and has been approved for publication. The work represents the views of the author and does not necessarily reflect the views of the agency.*

Introduction

Are you healthy? Is your environment healthy for you? What does it mean to be healthy? Does it mean you're not achy, or that you are not visibly sick? The World Health Organization defines health broadly as a state of physical, mental, and social well-being. This means that being healthy is more than just living a long life, but living a life that is productive and free from ailments that reduce your potential to live enjoyably.

In this chapter, we will focus on the aspects of our environment that affect our own health. These environmental influences can be classified into three categories: **infectious agents**, **chemical agents**, and **physical agents**. Infectious agents are living microorganisms that cause disease. Physical agents include things such as radiation and microwaves and other electromagnetic fields. Some of the effects of radiation on health are discussed elsewhere in this book. Chemical Agents is a broad category of metals and chemicals that contaminate our water, air, food supplies, and residential environments.

So how important are each of these in our lives? We are impacted by each of them every day, and we have been throughout the history of our evolution! Since before birth, our bodies have been assaulted by environmental invaders. We are bombarded by damaging rays from the sun, been selected as the perfect home by bacteria, and your body, a finely tuned biochemical machine, has been disturbed by foreign chemicals. Many of these environmental insults over time lead to what we know as aging, a natural process that we cannot avoid. Even if we could live in a bubble, cut off from as much of the environment as possible, the very oxygen we breathe causes damage to cells through a process called oxidation. Besides, what quality of life would living in such an environment provide? In the field of environmental health, we strive to reduce the influences of our environment on our quality and length of life to the extent possible, realizing that we cannot eliminate environmental impacts altogether. In the modern world, when we are producing new technologies and chemicals at a rapid pace, we are putting new things into our environment all the time. Some of these things exist in small amounts and may have subtle effects on our health. It is this subtlety that makes some of the problems in environmental health challenging and their scientific outcome often controversial. Here, we will begin to explore some of the concepts used in the study of environmental health and will examine in detail two case studies of environmental health issues that have arisen in Nevada. First, we will more clearly define the different types of environmental agents that affect our health.

Infectious Agents

Infectious agents can reproduce inside living organisms and escape to infect other living organisms. Most of us have experience with infectious agents on a regular basis. A person around us has a cold and sneezes, releasing cold viruses into the air. We breathe in these viruses, and if the conditions are right, they can reproduce inside our bodies, producing more viruses. The responses in our bodies result in the symptoms we associate with a cold. Often, infectious agents may attack our bodies without our knowledge, because our immune system destroys these agents before they can reproduce and cause damage.

Throughout human history, infectious agents have been a leading cause of death (mortality) and illness (morbidity). Smallpox, a highly contagious disease caused by a virus, is characterized initially by fever, headache, muscle aches, and finally a characteristic skin rash. Often, toxicity produced by the virus itself or secondary infections resulting from general body weakness have led to death. Viruses are changing all the time to protect themselves from hazards in their environment. When a virus changes, the new slightly modified version is called a new *strain*. Depending upon the strain of the smallpox virus, the mortality following smallpox infection varies widely between 10% and 62% (Fenner, 1988). This type of variability in response to infectious and toxic agents is an important factor in many aspects of environmental science and environmental health.

Viral infections, including not only smallpox but also the common cold and what we call the flu[1], cannot be "cured" by medical intervention such as antibiotics. When we get a cold, we can take medicine to reduce the symptoms, but the cold must run its course. Usually, within five to seven days our bodies are able to isolate and destroy the cold virus. Some viral infections may be prevented by vaccination. A vaccine contains a deactivated or weakened form of the virus. When it is injected into the human body, it looks enough like a live virus to cause the body to build up its defense mechanisms against the virus. With strong defenses, if the real virus infects the body, this virus can be destroyed before it reproduces significantly and causes harm. Due to effective and widespread vaccination programs, smallpox has not been a public health hazard for over 30 years.

Eradicating smallpox was no small feat. Records indicate that it had been infecting human populations since before 1100 AD (Fenner, 1988). Its ability to spread rapidly in human populations earned it a legendary place in history. The transmission of infectious agents such as smallpox is linked to conditions in our environment. Many of the issues we study in environmental science such as population trends and water and air cleanliness, and even climate are relevant.

The transmission of infectious agents is directly tied to population density and the age structure of the population. In countries or regions with more people in a given area, the possibility of transmission from person to person will be higher. The age structure of the population also affects the fatality rate. Young children and the elderly are often most susceptible to infectious agents, and the infections may be more severe. In many African countries, dense populations with a high proportion of young people allow for rapid transmission of infectious diseases. Political and economic factors also play a role in a country's ability to provide health services and protect its population against many diseases. For example, a less severe form of smallpox was endemic in the United States after the more deadly form had been eradicated. The death rate due to this minor form of smallpox was less than 2% (Fenner, 1988). This was considered a major public health risk in the United States and warranted sufficient attention to get it eradicated as well. At the same time, developing countries in Asia and Africa considered the minor form to be a nuisance and not worth spending their public health budgets worrying about as they battled the more serious form of the disease.

1. We typically refer to a variety of viruses such as stomach viruses as the "flu". The actual influenza virus causes respiratory symptoms similar to the common cold along with fever and aching.

Infectious Disease Worries in Nevada

Modern jet travel rapidly brings people to Nevada from around the world. While smallpox is no longer a significant risk, there are other viral infections common in other places in the world to which we in the United States normally have little exposure or immunity. It is difficult to pinpoint particular infectious disease worries that are unique to Nevada. However, our unique distribution of population does pose some interesting challenges. While the population density of Nevada is extremely low, there is relatively high population density in Washoe and Clark counties. Clark County, in particular, accounts for a majority of the state's population. Because Nevada relies heavily on tourism for its economic well-being, it is important that we are cognizant of infectious disease and public health to protect our visitors and residents alike.

In the Las Vegas area, 85% of the drinking water is supplied from Lake Mead via the Colorado River. It is extremely important to protect the cleanliness of this resource. In 1994, in Clark County, 78 people contracted illnesses related to the bacteria *Cryptosporidium*, which was found in the city drinking water supply. Cryptosporidosis is the disease caused by cryptosporidium, a protozoan, or single-celled parasitic pathogen that lives in the intestines of animals and people. Contamination of water can occur through feces if the water is not properly treated before drinking. According to the *Las Vegas Review-Journal,* 32 of those infected with *Cryptosporidium* died from this outbreak. Many of them had immune systems already weakened by HIV, the virus that causes AIDS (Rogers, 1997). The source of the *Cryptosporidium* in the drinking water was never determined, and, according to the Centers for Disease Control and Prevention (CDC), the fact that the illnesses were related to drinking water wasn't determined until after the outbreak had been contained (Kramer, 1996). The Las Vegas outbreak was minor compared to another outbreak the same year, in which over 400,000 people in Milwaukee, Wisconsin, became ill.

Chemical Agents

Chemical agents include a wide variety of things, such as metals and other minerals from the earth, more than 3,000 chemicals produced in high volume for industrial use, chemical additives in our foods, pesticides, and other chemicals in our air and water. Chemical agents can also include a group called **toxins**, chemicals produced in nature, often for defensive purposes by living organisms. **Toxicology** is the study of the adverse effects these chemicals have on humans, animals, or ecosystems. In discussing environmental health, we are particularly interested in human toxicology and how this is related to potential health risks resulting from exposure to chemicals in the environment.

Toxicologists often refer to exposure and toxicity in the context of different time frames: acute and chronic. Acute effects occur following a single high-dose exposure to a chemical. Acute exposures are usually accidental. For example, a child drinks from a bottle containing cleaning fluid or a spill in the workplace causes high exposure to a chemical. In this situation, effects are often immediate. Uncertainty regarding the types of effects, their severity, and expected response is often low. We know the symptoms of poisonings, and physicians typically know the best course of action for treating them. When we talk about acute effects, we may be concerned with the potential for **lethality**, or death, resulting from the exposure.

Toxicologists explore the relative acute toxicity of different chemicals, often using experiments on animal species that may or may not closely resemble humans. Rats or mice are often used because they are easy to handle in large numbers and because relatively few ethical issues arise over the use of these animals. Even though these rodents aren't as similar to humans as are some other animals, the differences in their physiology and biochemistry have been well studied. Scientists typically know where the differences lie and can account for those in their **extrapolation** of the effects to humans. One obvious difference between rodents and humans is their size. Since the toxicity of something is dependent upon its dose, we can all understand somewhat intuitively that the toxic dose in a larger creature will be greater than in a smaller creature. When we prescribe drugs, the recommended doses in small children is always less than the dose for adults. For this reason, just as when we measure the effectiveness of drugs, toxicity values must take into account differences in body weight. **Lethal doses** of a chemical are therefore presented relative to body weight (i.e., per kilogram). Because of differences in genetics and physiology between individual animals, the same dose of a chemical will not cause the exact same response (effect) in all test animals. Small doses of compounds will not have any effect. As the dose increases, a larger percentage of animals will be affected and some may die. At a high enough dose of a chemical, all tested animals will die. The dose level that results in death to 50% of the test animals in a study is known as the **LD_{50}**. This measure is used to explain the relative toxicity of different poisonous chemicals. The LD_{50} for several common chemicals are listed in Table 6.1. These data can be obtained from Material Safety Data Sheets (MSDS), which are kept by any organization, (including universities and companies) where chemicals are used. LD_{50} values can exist for many things from unusual chemicals to common nutrients and pharmaceuticals. This is another example of how the dose determines the toxicity.

The LD_{50} measures one particular effect: death. Certainly, other effects also result from exposure, but many of these are subtler effects that may occur after **chronic**, or long-term, often low-level exposure. Some chemicals such as dioxins and polychlorinated biphenyls (PCBs) are not readily broken down and excreted by the body. They remain there for a long time. After long-term low-level exposure, they can **bioaccumulate** and could potentially build up to toxic levels in the body. Some of the effects of chronic and acute low-level exposure to chemicals are discussed below.

What is important to realize in these discussions is that almost all known chemicals, both synthetic and natural, can be harmful. The critical factor that determines whether something is harmful, harmless, or helpful is the *dose*, the amount of a substance that a person has inhaled, ingested, or absorbed through the skin. Even nutrients and foods can be harmful. The difference between a nutrient and a poison is the amount of the substance that is harmful. Vitamins and minerals are chemicals that our bodies require and utilize as nutrients. These are naturally occurring organic and inorganic molecules that are the building blocks of our cells. Some function as vital enzymes in chemical reactions in our bodies. However, these chemicals can be poisonous and can cause death if ingested in large quantities. Some vitamins, such as vitamin C, are water-soluble. If an excess amount is ingested, it is readily excreted in urine. Other vitamins, such as vitamin A, are stored in fat. These fat-soluble vitamins have the potential to build up to toxic levels in the body if excess vitamins are ingested. Chromium and selenium are minerals, inorganic metals found in the earth, and are a necessary part of the diet. In large doses, these two metals can cause brain damage. For example, oxygen, an element that is so necessary for life, can also be toxic to cells. Even oxygen can be toxic. Lung cells bathed in pure oxygen will die, but they function quite well in an environment of approximately 15-20% oxygen, the amount that exists in our atmosphere.

Table 6.1. Lethal Dose (LD_{50} mg/kg) of Various Common Substances

Substance	**Route – Animal**	**LD_{50} (mg/kg)**
Sodium Chloride (table salt)	Oral – rat	3000
Sodium Perchlorate	Oral – rat	2100
Ethanol (grain alcohol)	Oral – rat	2080
Acetaminophen (pain killer, Tylenol)	Oral – rat	1944
Acetaminophen (pain killer, Tylenol)	Oral – human	143 (LD_{LO})
Caffeine	Oral – dog	230
Caffeine	Oral – rat	192
Nicotine	Oral – rat	50
Sodium Cyanide	Oral – rat	6.4

When a product label says "nontoxic", the manufacturer really means the chemicals in this product are present in small enough amounts and that their toxicity is low enough that they are not likely to cause harm to people. Often times advertisements for pest-control products say that a product is nontoxic to people and pets. All pesticides are poisons, in that they have no nutritive benefit to humans or animals. They are designed to cause death to certain organisms. In designing modern pesticides, the goal has been to develop chemicals which affect the physiology of the pest but that have less of the toxic effect on humans and other non-target species. This is called *specificity.* Most of the time, the physiology between different animals is similar enough that finding a chemical that is very toxic to one species but very nontoxic to all other species is difficult.

Over the years, our bodies have evolved mechanisms to handle and process natural chemicals at the levels to which we are typically exposed in the environment. Chemicals are chemically changed in our bodies to create and excrete harmless substances. Alcohol, for example, is detoxified in our liver. Chemicals that humans produce and have developed as part of our industrial processes over the last several hundred years are more of an issue for our bodies, because they have not evolved in the presence of these chemicals. It is fortunate that environmental health officials have become aware of possible problems, and our exposure to many dangerous chemicals in the air and water has been reduced in recent years, especially in the United States and other developed countries. Our ability to detect these chemicals continues to improve, so that we are now able to measure very small amounts of chemicals in the environment. For many chemicals, we can measure levels in air and water in the part-per-billion range (ppb). One ppb is equivalent to approximately one ounce of liquid in about eight million gallons of water.

Chemicals that can exert toxic effects on the nervous system are called **neurotoxins**. Tetrachloroethylene, a chemical used to dry clean clothes, has anesthetic properties if inhaled. Some acute effects of tetrachloroethylene exposure include drowsiness and the inability to concentrate, the same effects we desire in the anesthetic drugs used by physicians during surgery. Many insecticides are neurotoxins. Organophosphate insecticides such as diazinon and chlorpyrifos block the action of acetylcholinesterase, a chemical that controls signal transmission between nerve cells and muscles.

Most cells in our bodies reproduce on a regular basis. We all know that new hair and fingernail cells are being generated all the time, as are skin cells. Cells reproduce inside our organs to replace cells that have died or been damaged, and they do this at a well-defined rate. During growth and development, growth does not occur haphazardly When our organs are not growing, as in adults, the cells reproduce to maintain the current size, shape, and function of the organ. The rate at which cells reproduce is controlled by physical signals between adjacent cells as well as genetic programming in cells by deoxyribonucleic acid, or DNA. Sometimes, this well-controlled process can malfunction, and cells not only reproduce out of control, but they also no longer express the genetic characteristics of whatever organ they were intended to be part of. When this happens, we call this process of uncontrolled cell reproduction, cancer. It is also known that certain chemicals called **carcinogens** can be involved in the process of cancer. Cancer rates have risen in most industrialized countries during the past century, and many scientists suspect that carcinogens in the environment may be partly responsible for this. During the past 100 years, there have also been lifestyle changes (smoking, diet, etc.) that could be implicated. We are also living longer today than we ever have before. We know that genetic alterations involved in beginning the process of cancer occur regularly, but generally, our cells "catch" this and stop the process before it progresses. The longer we live, the greater the likelihood that at some point the regulatory mechanisms in the cells may not catch a precancerous step and it may be permitted to progress. Whether many cancers can be attributed to chemicals in the environment or not, scientists have demonstrated in animal studies that certain chemical exposures as well as some physical exposures, such as ionizing radiation, can cause cancers. Many of these exposures, such as the radiation in sunlight, cannot and should not be eliminated completely. However, it is prudent public health policy to try to reduce the population's exposure to the most potent known carcinogens as much as possible.

Some chemicals are toxic to developing organisms if a mother is exposed during pregnancy.

These **teratogens** may not affect adults but can cause birth defects in developing embryos and fetuses. It is now well known that drinking alcohol during pregnancy can lead to a variety of symptoms in children from physical malformations to slowing of mental development. Perhaps one of the most famous cases of intentional (although unknown at the time) teratogen exposure was the regular prescription of a drug called Thalidomide, a sedative first used in the 1950's to relieve morning sickness in pregnant women (Moeller, 1997). It ended up causing a large number of children to be born without arms or legs.

Immunotoxins are chemicals that are known to alter the function of the immune system. Many of us have experienced allergies to certain things in our environment or certain foods or drugs. An allergic response is a heightened immune response – that is, our body recognizes a foreign substance and activates the immune system to try to rid the body of it. As in anything else, different people have different sensitivities to these immune modulating effects of allergens. Some people have allergic responses to chemicals in pet hair, silver, plastics, to name just a few.

Case Study: Rocket-Fuel Oxidizer in Southern Nevada Drinking Water

Sodium perchlorate is a chemical used in the production of rocket fuel, an industry that was especially prominent in Henderson, Nevada (Clark County) during and immediately following World War II. The LD_{50} for sodium perchlorate (also called simply perchlorate) is listed in Table 6.1. It has been found in high concentration in soil samples taken near industrial facilities in Henderson and has been detected in relatively high concentrations in The Las Vegas Wash, the stream that carries waste and flood water from the Las Vegas Valley through Henderson and southeast to Las Vegas Bay in Lake Mead. Perchlorate has been detected in water supplies in at least 18 other states, including Arizona and California, usually in areas near industrial sites.

As indicated in Table 6.1, it would take a significant amount of sodium perchlorate to cause death. Therefore, acute toxicity resulting from environmental exposure to low concentrations is not a concern.

Not much is known about the chronic effects of low-level exposure to perchlorate because it has never been an environmental issue until around 1998, when it was detected in the drinking water

Physicians have used perchlorate to treat certain thyroid disorders. The thyroid gland is part of the endocrine system. It produces hormones that affect metabolism, the rate at which the body burns fuel and produces energy. If metabolism is high, a person is more alert, consumes more food (energy), and produces more body heat as a result. However, these people also typically have trouble focusing and concentrating and may have excessive energy and no outlet for it. If metabolism is low, a person may be more lethargic, will not utilize as much food energy (although a person may still consume many calories, resulting in obesity), and the body temperature may be below normal. Perchlorate blocks the function of the thyroid by preventing its uptake of iodine, a critical ingredient in many of the hormones produced by the thyroid. Clinically, physicians have used perchlorate to treat hyperthyroidism, the condition in which the thyroid is overactive, leading to high metabolism. It is not clear what chronic dosing of low-levels of perchlorate from the environment may do in healthy individuals.

Chronic toxicity is most often studied using laboratory animals. Animals such as rats have a much shorter life span than humans do, so doing a lifetime study in animals takes less time than doing the same study in humans. Lifetime studies can examine the effects of a chemical during different stages of life or after a full lifetime of exposure in the animal. It has been suspected that any toxicity of perchlorate may be due to its effects upon the thyroid gland. Long-term animal studies have indicated that by acting like a growth hormone, exposure to low-levels of perchlorate can increase the weight of the thyroid and cause individual thyroid cells to become larger (Siglin, 2000; Modric, 1999). The chemical seems to have no other noticeable effects on the animals. What is not known is if the abnormal size of the thyroid and shape of the thyroid cells results in any difference in thyroid function that could lead to health complications.

Using toxicity studies in animals to understand human safety and risk from chemicals has some difficulties and uncertainty. For example, chemicals may not have the same effects on rats and mice as they do on humans. Therefore, scientists attempt to understand as accurately as possible the biological effects of the chemical in the animals, down to the cellular level. If they know this and if they understand precisely how animals are different from humans in these specific aspects of their biology, then the scientists can have more certainty on the effects of the chemicals in humans.

In addition to using animal studies, we can try to understand the health effects of chemicals in the environment through **epidemiology**, which examines the health of actual human populations. Epidemiologists look for relationships between factors in the environment and a particular health outcome. For example, several epidemiology studies have indicated that people living in cities with more air pollution may have a greater risk of lung disease than those living in areas with cleaner air. In the case of perchlorate, scientists have examined medical records for people living in Southern Nevada who receive over 80% of their drinking water from Lake Mead and compared these records to those of people living in Washoe County, where perchlorate has not been found in the water supply. In examining the records, the scientists were looking specifically for thyroid disorders and thyroid cancers. Over a two-year period, 1997 and 1998, researchers found no higher incidence of any thyroid disorders in Clark County relative to Washoe County (Li, 2001). The body closely regulates the activity of the thyroid. A small gland near the base of the brain called the hypothalamus accomplishes this regulation by secreting a hormone called Thyroid Stimulating Hormone (TSH). To increase the activity of the thyroid and the overall rate of metabolism in the body, the hypothalamus secretes more TSH. To slow the thyroid (and the overall metabolism) down, the hypothalamus secretes less TSH. When perchlorate blocks the normal function of the thyroid, in effect slowing it down, one effect that can be seen clinically is increased levels of TSH, as the body tries to compensate and speed up the thyroid. It is as if you are driving a car, and all of a sudden, something happens to your engine. It sputters and the car begins to slow down. Your natural inclination is to try to speed the car up by stepping on the gas pedal. TSH acts like the gas pedal for the thyroid gland, in effect encouraging it to go faster.

The association between levels of TSH in the blood and exposure to environmental levels of perchlorate is unclear. Evidence supporting an association has been presented by a group of epidemiologists from Arizona who studied TSH levels in two groups of newborns in Arizona: newborns from cities where drinking water comes from the lower Colorado River and contains perchlorate, and newborns from cities whose water does not contain the chemical (Brechner, 2000). While another study carried out in Chile, which is known to have a high level of natural perchlorate in its soil and in some of its groundwater supplies, found an opposite relationship. In the Chile study, newborns in cities with high perchlorate levels in the water actually had lower levels of TSH in their blood than children born in cities with low perchlorate (Crump, 2000). The highest levels of perchlorate detected in the drinking water in one of the Chilean cities was 120 micrograms/Liter, while in the Colorado River and Lake Mead in Arizona and Nevada, perchlorate has been detected at levels no higher than 18 micrograms/Liter.

Remember that the blood levels of TSH are not any direct indicator of thyroid function or abnormal thyroid. Elevated levels simply indicate that the hypothalamus is trying to "speed up" the thyroid gland. In another study, nine adults intentionally took doses of 10 mg (10,000 micrograms) of perchlorate per day for 14 days (Lawrence, 2000). The researchers investigated several aspects of thyroid function. They reported that the circulating levels of TSH in the blood of these individuals were unchanged. However, they also noticed that the uptake of iodine by the thyroid was reduced. Iodine is a critical ingredient that the thyroid uses to perform its functions of regulating metabolism in the body. The next unanswered question is: at how low a level of perchlorate is the iodine uptake by the thyroid reduced?

In early 2002, the U.S. Environmental Protection Agency (EPA) released its draft health assessment for perchlorate recommending that the concentrations of the chemical in drinking water should not exceed 1 microgram/L, a value considerably below the 18 micrograms/L that California had been using as the action level for shutting down municipal drinking water wells. The EPA set its values based on animal studies completed over the last several years that indicate that the effect of perchlorate on the thyroid gland can be particularly damaging to a developing fetus. The EPA acknowledges that even at several micrograms/L, the risk of harm from perchlorate is low but still represents a risk, and that understanding the health effects of perchlorate require more research (Danelski, 2002). Before the EPA recommendation can be enforced upon municipal water supplies, the President must review it and sign it. If that occurs, some drinking water supplies in California and Nevada as well as in 18 other states will have perchlorate levels at several times the level deemed safe. These water suppliers must then make efforts to reduce the perchlorate concentrations to below the EPA recommended limit.

Case Study: The Leukemia "Cluster" in Fallon

As humans, we are constantly searching for causes of things. We like to understand how things work and to believe that we can attach a reason to things that occur.

In 1999, two children in the small town of Fallon, Nevada, were diagnosed with leukemia, a type of cancer affecting white blood cells. These specialized blood cells fight infection in our bodies. When these cells become cancerous, they no longer function in their normal way, leaving the body open to serious and life-threatening diseases. The two most prevalent types of leukemia are Acute Lymphocytic Leukemia (ALL) affecting predominantly children, and Acute Myelogenous Leukemia (AML), affecting primarily adults. The names of these diseases reflect the specific white blood cell type involved.

Both 1999 cases in Fallon were ALL cases. This did not gain a particular amount of attention. However, in 2000 when nine additional cases of ALL in children were identified in this small town of 8200 people, it attracted the attention of local health officials and the media, demanding an investigation of this "cluster" of cancers. By the end of 2001, there were 14 confirmed cases of Leukemia (13 ALL and one case of AML) in children with ties to Fallon. It should be noted that this number includes children that had lived in Fallon during their childhood but may no longer live there. Because of the Fallon Naval Air Field, the town has a transient, military population. One former Fallon resident, a woman, died in 2001 in Pennsylvania at the age of 21.

Is the case of Fallon unusual? How many cases of leukemia would one expect in a county the size of Churchill County, Nevada? These questions take us to the study of risk, a branch of applied statistics. The statistics can tell us how many cases of leukemia one can *expect* in Fallon, or more generally, Churchill County. If there are more cases of leukemia than what would be expected, the statistics can also tell us the probability of whether such a cluster occurred by random chance. National statistics suggest an average rate of ALL of three children per 100,000 people in the population (Schinazi, 2000). According to census figures, the estimated population of Churchill County on July 1, 1999 was 23,405. The population is relatively stable, growing at a modest 1.1% annually (www.census.gov). With this small population, the national statistics suggest that one could expect to see one case of ALL in Churchill County approximately every five years! Clearly, the confirmed 14 cases are significantly above the expected value. How unlikely is it, though? Also, can any environmental influences that may be involved in this higher than expected number of leukemia cases be isolated?

Epidemiologists investigate the possible causes of such "clusters" using a study design called **Case-Control**. In this type of study, the environment of the people with the disease is examined and compared to an environment in which people do not have the disease (controls) to see what, if anything, the individuals with the disease may have been exposed to in the past with higher frequency or duration than the control individuals. Case-Control studies were used to establish the evidence that smokers have as much as a 30 times greater risk of lung cancer than non-smokers (Moeller, 1997).

Most studies in environmental epidemiology are not this clear. Cases of environmental contamination and resulting disease have been dramatized by popular movies such as *Erin Brockovich* and *A Civil Action*. In these movies, the crusaders, representing people whose children have been struck with horrible diseases, go after and get the big corporate entities responsible for causing the disease and death. These two movies were based on true stories – very high profile cases resulting in large jury awards to the families affected by disease. This means that the jury was convinced that the companies were responsible, even in the face of what was likely significant scientific uncertainty. So, legally, these cases were won. But epidemiology and science rarely find absolute answers in cases like this (Gawande, 1999). This brings us back to the case of Fallon. Certainly, it is shocking that 14 cases of cancer have been linked to Fallon, Nevada over the past four years. Certainly, it would be tragic if there were a single environmental factor causing these cancers and nothing was done to reduce the risk. In a 2001 hearing before the Senate Committee on Environment and Public Works, Senator Harry Reid of Nevada said: "I can think of nothing more heartbreaking than a child suffering with a serious health condition, and nothing more frustrating than not knowing the cause. Yes, we are facing a highly complex situation, and we do pretend to think there are easy answers. But, I give you my commitment to support investigations and efforts to address the causes of the childhood leukemia cluster and other health threats facing this community [Fallon] or any other" (Congressional Testimony, 4/12/2001).

Another member of the committee, Senator Hillary Clinton from New York, also faced similar, unexplained cancer pockets in her state. Possible culprits in the Fallon area ranged from radiation involved in nuclear testing to materials used by the U.S. Navy at the Fallon Naval Air Station to a high level of arsenic found in the groundwater supplying Fallon residents with their drinking water. The arsenic became the primary chemical of interest to the media and the public, in part because the media attention that the Fallon leukemia cases received coincided with a national debate over the permissible level of arsenic to be allowed in drinking water by the U.S. Environmental Protection Agency (EPA).

Arsenic, a naturally occurring metal, can enter groundwater by leaching out of rocks and soils or by direct contamination from an industrial source. In Fallon, as in many places in the west, there are natural sources of arsenic which result in water supplies containing levels of this chemical that approach the current maximum limit set by the U.S. EPA of 50 parts per billion (ppb).

At acute, high doses, arsenic is a well-known poison of epic proportion. The LD_{50} in humans is between 1 and 4 mg/kg. This means that a 150 pound human would probably need to ingest only between 70 and 280 mg (between 0.002 and 0.009 ounces) to cause death! However, the levels in the environment are typically millions of times lower than this. At lower doses, the effects have been more uncertain. Of course, we now know the dose determines the toxicity, and even with something as toxic as arsenic, there has been speculation that arsenic in small doses is an essential nutrient (U.S. EPA, 2000).

In January 2001, the U.S. EPA revised its standard for arsenic, lowering the Maximum Contamincation Limit (MCL) permissible in municipal water supplies from 50 ppb to 10 ppb by the year 2006. At levels of 50 ppb in water, a 150 pound person would have to drink 364 gallons of water per day to reach a dose of 1 mg/Kg, the low end of the LD_{50}. With water levels at the new EPA standard, the amount of water needed would be 5-fold less, or approximately 73 gallons of water. Levels of arsenic in Fallon have been detected as high as 100 ppb, but average at around 15-20 ppb, such that it does not require cleanup under the old standard but would require cleanup under the new one. The arsenic standard setting was an extremely contentious issue nationwide, with proponents saying that arsenic was clearly a carcinogen, a chemical that causes cancer, and that to allow higher levels in drinking water would be unacceptable. Opponents of the new standard said that it would immediately put many small water systems in the West in noncompliance with the law, forcing them to spend millions of dollars to upgrade treatment systems to remove the contaminant. These people argued that if the federal government is to make such strict requirements, it should provide the money to upgrade the water systems. In fact, legislators, including Senators Harry Reid (D) and John Ensign (R) of Nevada used the proposed standard to justify regulation giving $750 million to small communities (towns of less than 10,000 people) to upgrade their water systems (Dorsey, 2001). In Fallon, where arsenic levels have been measured as high as 100 ppb, or more than 10 times the proposed MCL, would require approximately $6.7 million for the equipment to treat the water and bring it into conformation with the proposed standard (Associated Press, 2001).

Several political and scientific issues are involved in this case. The scientific issues are: there has been a high rate of, in particular, acute lymphocytic leukemia (ALL) observed in the Churchill County area and specifically, the town of Fallon, Nevada since 1997. Another issue is that residents of Fallon drink water that contains levels of arsenic that are higher than some other, but probably not all, US residents. The interpretation of the scientific evidence is influenced by political factors. Amidst all of these issues, neither science nor politics has determined whether there is a relationship between arsenic and the cases of ALL observed in Fallon. In fact, many public health officials are skeptical that arsenic could be involved in the leukemia cases.

The relationship between arsenic and some types of cancer is becoming clearer from multiple animal and epidemiology studies. Several animal studies and Case-Control studies of human populations exposed to arsenic in drinking water provide strong evidence that indeed arsenic is a carcinogen but leukemia has not been among the types of cancers seen in animals and humans. A study of over 40,000 people in Taiwan revealed increasing rates of skin cancer possibly linked with increasing arsenic level in village water supplies. The water supplies contained arsenic levels from 0 to greater than 60 ppb. Other studies of arsenic exposure leading to skin cancer in animals and humans led the EPA to classify arsenic as a carcinogen based on the increased risk of skin cancer. If arsenic at levels as low as 60 ppb could be implicated in increased skin cancer risks, even if that increase is small, this is, perhaps, an argument to lower the arsenic standard for drinking water from 50 ppb to 10 ppb. The risk can never be certain from a single study. A recent study in a small Utah town where arsenic concentrations in the water averaged between 18 and 191 ppb revealed no statistically significant difference in cancer rates in this town relative to the rest of the U.S. population (Lewis *et al.*, 1999). Other types of cancers have been associated with arsenic exposure, but these are primarily cancers of the internal organs like kidney, lung, and bladder (U.S. EPA, 2000). These relationships have been noted in animals but have not bee observed in humans.

Arsenic in drinking water cannot be scientifically related to the cases of leukemia that have arisen in Fallon. Despite assertions of this from local, state, and federal public health officials, a concerned public cannot help but be concerned about the arsenic or whatever environmental contaminant may be causing what is believed to be an unnaturally high rate of leukemia in Fallon. This affects the local economy in many ways. Navy personnel threaten lawsuits if assigned to the Naval Air Station there, and the town has to deal with the fallout of people wanting to leave or not wanting to move there.

Many scientists and public health officials believe it will be very difficult to find an environmental cause for the leukemia cases in Fallon, but the investigations by the Nevada Division of Health and the U.S. Centers for Disease Control and Prevention are ongoing. In the past, environmental epidemiology has detected specific links between environmental contaminants and disease. However, the most convincing studies are those in which a relationship between increasing dose and increasing response rate can be shown. In the Taiwan study described earlier, there was evidence of increasing rates of skin cancer with increasing doses of arsenic in the drinking water. Chemicals in water are particularly difficult, because many residents of Fallon, for example, drink bottled water and are therefore be exposed to fewer contaminants from the city's drinking water supplies.

Some scientists are more certain that an environmental link for the leukemia cases in Fallon can be found. Professors Mark Witten and David Harris from the University of Arizona are interested in a leukemia cluster in Sierra Vista, a small town south of Tucson, Arizona. This town is home to an Army base and airfield. The presence of military air operations in this town and in Fallon, Nevada, have given the Arizona researchers reason to speculate that some component of jet fuel released into the air or into the drinking water may be associated with the leukemia cases. These researchers admit they have no scientific proof that the jet fuel causes cancer in humans, and Arizona state health officials warn the public not to jump to any conclusions (*Las Vegas Review Journal*, 2002). Some supporting evidence may suggest that the link between leukemia and jet fuel is possible. In one study, the activities of 640 people with ALL were compared with those who did not have ALL. This design, as you may recall, is called a **Case-Control Study**. These researchers found that children who frequently were exposed to model building glue and who participated in artwork using petroleum-based solvents had a higher risk of developing ALL (Freedman, 2001). Naturally, there are other explanations, and problems are raised with these types of studies, but they create the premise for a possible link between ALL and petroleum-based products, such as jet fuel.

The Statistical Problem

While it may seem unlikely, it is quite possible that cancer rates, although higher than average, could have occurred by chance. On a roulette wheel, there are an equal number of red and black slots. There is an almost 50% chance, or one in two, that the ball will land in either a red or black slot (there is usually one or two zeroes which are neither red nor black, making the probability slightly less than 50%). If you have been in a casino, you know that it is possible to see R-R-R-R-R-R appear on the roulette wheel. Each spin seems to hit red. Does this mean the wheel is rigged? Does this mean that the ball is more likely to land on black next time? The answer to both questions is no. There is still approximately a 50% chance that on the next spin, the ball will land on red. Cancers are rare events, but not that rare. One child in 33,000 will get ALL every year. In the year 2000, the Nevada population was estimated at 2,018,723 from the U.S. Census (the population increased to 2,106,074 in 2001). In Nevada and nationwide, the percentage of the population under the age of 20 (ages 0 – 19) is approximately 28%, according to U.S. census figures. This means there were approximately 589,700 teenagers in Nevada in 2000. If one in every 33,000 will get ALL, one can divide the total population by 33,000 to get the number of ALL cases expected each year in Nevada. That number is just over 17 cases per year. It is expected that the cases will be evenly distributed throughout the population. The key word in both of these sentences is "expected". These are averages. The statistics tell us that if we look over long periods of time this is what we should see. There are, of course, natural variations in that. Depending upon how far the number of cases is from the average value, statistics can tell us what the likelihood is that this number of cases arose by random chance – that is, natural variability around the average number of cases. The farther away from the average, the less likely it happened by random chance.

The questions are, how many cases occurred in Nevada, and how many occurred in Fallon (or Churchill County)? The State of Nevada, in cooperation with all of the medical centers in the state and the U.S. Centers for Disease Control and Prevention, maintains the Nevada Cancer Registry, a database containing valuable information on all of the diagnoses of cancer in the state. Table 6.2 shows the leukemia cases as stored in the Nevada Cancer Registry broken down by year. ALL and AML are listed separately, and all other leukemia types are lumped together. Diagnoses in Churchill County are pulled out of the state totals in this table. If you look at the state totals from 1995 – 2000, you will see that in children ages 0—19, there were fewer cases of ALL than expected in 1995 – 1998, with no more than 11 cases each year. In 1999, there were 18 ALL cases and in 2000, the preliminary results indicate there were 19 ALL cases. Based on the expected value of 17 that we calculated above, these don't seem very unusual. Next, look at the cases for Churchill County shown in Table 6.2. There were no cases reported at all in this database until the year 2000, when there were three.

You will recall that this seems contradictory to what we discussed earlier in the chapter, and with what has been reported by the media. Table 6.3 highlights the difference. This table, produced by the people in the State Department of Health who are investigating the cancer cluster, reports cases in a different way than is reported in the Nevada Cancer Registry. The NCR reports cases diagnosed by residents of Nevada and residents of the particular county at the time of diagnosis. What is reported to the media is the data in Table 6.3 that has been tabulated after the cancer cluster was identified and researchers began investigating it. These cases are people who were diagnosed during a given year and had lived in Fallon sometime previously.

This method of reporting is what some epidemiologists call the "Texas-sharpshooter fallacy" (Gawande, 1999). An unscrupulous sharpshooter may shoot at the barn wall first and then draw a bulls-eye around the bullet later. In a similar manner, when we begin to notice a few excess cancer cases in a city, we draw a cohort around them, define this population as special, and start to examine it with extra scrutiny, perhaps turning up extra cancer cases in ways that ordinarily would not show up. In Fallon, there were no diagnoses of leukemia for people living in the area in the years 1995 – 1999. In 2000, four cases of leukemia were diagnosed in the Fallon area (three cases of ALL). This is a large increase in the number of diagnoses over what the town experienced previously and over what one would expect. This caught the attention of the local residents who made their

Table 6.2. Reported Leukemia Cases in Children (0-19 years) by County of Residence at Diagnosis and Type of Leukemia, Nevada Residents, 1995 – 2000

Year of Diagnosis	Type of Leukemia			Total
	Lymphocytic Leukemias (ALL)	Myeloid Leukemias (AML)	Other Leukemias	
Churchill County				
1995	0	0	0	0
1996	0	0	0	0
1997	0	0	0	0
1998	0	0	0	0
1999	0	0	0	0
2000*	3	0	1	4
A. Nevada				
B. Total				
1995	10	4	2	16
1996	11	2	1	14
1997	11	3	2	16
1998	11	5	2	18
1999	18	5	3	26
2000*	19	2	4	25

*Data may not be complete and is preliminary as of 12/20/2001.
Source: Nevada Division of Health.

concerns known to the State Department of Health. However, if you look at the Nevada state numbers during the same year, they didn't change much from the previous year. It is possible that the four cases of leukemia diagnosed in Fallon in 2000 were like the R-R-R-R on a roulette wheel. It is statistically unlikely that they would occur together, but possible. As part of their research efforts, public health officials establish a cohort that includes all children who have lived in Fallon in the past and have been diagnosed with cancer during a particular year (giving rise to Table 6.3). Like the sharpshooter who draws a bulls-eye around the bullet, this may be a bit unfair statistically, especially in an area like Fallon with a transient military population. While we compute the statistics based on the current population numbers in Nevada and the numbers currently in Churchill County, one begins to include leukemia cases that are no longer residents of the state and should technically be contributing to cancer statistics elsewhere.

The investigation continues in Fallon, and there may be an environmental reason that the cancer rates jumped in 2000, but it will be impossible to determine without more data over the next few years, and even then. These investigations are difficult. Where Hollywood and often our legal system like to find clear answers, epidemiologists are usually unable to find an environmental cause in most cancer cluster investigations (Gawande, 1999).

Despite the difficulties of environmental health research, the investigations continue. Like detectives trying to find an unknown killer, researchers continue gathering evidence to understand how agents we put in our environment may impact our health.

Table 6.3. Fallon Cancer Cluster Leukemia Cases (Ages 0 – 19)*

Year	Number of Cases
1997	1
1998	0
1999	2
2000	10
2001	2

* A case is defined as an individual having lived in Fallon at some time prior to the diagnosis.
Source: Nevada Division of Health.

Acknowledgment: The author would like to thank Dr. Peter Egeghy for his editorial and scientific review of this chapter.

References

Associated Press. 2001. Ridding water of arsenic costly. *Las Vegas Review-Journal*, February 3, 2001.

Brechner, R.J., G.D. Parkhurst, W.O. Humble, M.B. Brown, and W.H. Herman. 2000. Ammonium perchlorate contamination of Colorado River drinking water is associated with abnormal thyroid function in newborns in Arizona. *J. Occup. Environ. Med.* 42(8): 777-782.

Crump C, P. Michaud, R. Tellez, C. Reyes, G. Gonzales, E.L. Montgomery, K.S. Crump, G. Lobo, C. Becerra, and J.P. Gibbs. 2000. Does perchlorate in drinking water affect thyroid function in newborns or school-age children? *J. Occup. Environ. Med.* 42(6): 603-612.

Danelski D, J. Bowles. 2002. EPA raises red flag on inland water. *The Press-Enterprise*, Riverside CA, January 23, 2002.

Dorsey C. 2001. EPA withdraws arsenic regulation pending review. *Las Vegas Review Journal* March 21, 2001.

Fenner F, D.A. Henderson, I. Arita, Z. Jezek, and I.D. Ladnyi. 1988. *Smallpox and its eradication*, World Health Organization (ISBN: 92 4 156110 6).

Freedman D.M., P. Stewart, R.A. Kleinerman, S. Wacholder, E.E. Hatch, R.E. Tarone, L.L. Robison, and M.S. Linet. 2001. Household solvent exposures and childhood acute lymphoblastic leukemia. *Am J Public Health* 91(4): 564-567.

Gawande A. 1999. The cancer-cluster myth. *The New Yorker*. 8 February: 34-37.

Kramer MH, B.L. Herwaldt, R.H. Calderon, D.D. Juranek. 1996. "Surveillance for Waterborne Disease Outbreaks –United States, 1993-1994", CDC *MMWR Surveillance Reports*, 45(SS-1): 1-33, April 12, 1996.

Las Vegas Review Journal 2002. Arizona Leukemia Cluster to be studied. *January 28, 2002.*

Lawrence JE, S.H. Lamm, S. Pino, K. Richman, and L.E. Braverman. 2000. The effect of short-term low-dose perchlorate on various aspects of thyroid function. *Thyroid* 10(8): 659-63.

Lewis DR, J.W. Southwick, R. Ouellet-Hellstrom, J. Rench, and R.L.Calderon. 1999. Drinking Water Arsenic in Utah: A Cohort Mortality Study. *Environmental Health Perspectives*. 107(5):359-365.

Li FX, L. Squartsoff, and S.H. Lamm. 2001. Prevalence of Thyroid Diseases in Nevada Counties With Respect to Perchlorate in Drinking Water. *J Occup Environ Med* 43(7) 630-34.

Longnecker MP, M.A. Klebanoff, J.W. Brock, and H. Zhou. 2001. Polychlorinated biphenyl serum levels in pregnant subjects with diabetes. *Diabetes Care* 24(6): 1099-1101.

Modric T, K. Rajkumar, and L.J. Murphy. 1999. Thyroid gland function and growth in IGF binding protein-1 transgenic mice. *European Journal of Endocrinology* 141(2): 149-159.

Rogers, K. 1997. Officials confirm sewage release. *Las Vegas Review Journal,* May 20, 1997.

Moeller D.W. 1997. *Environmental Health.* Boston: Harvard University Press.

Schinazi R.B. 2000. The probability of a cancer cluster due to chance alone. *Statistics in Medicine* 19: 2195-2198.

Siglin JC, D.R. Mattie, D.E. Dodd, P.K. Hildebrandt, and W.H. Baker. 2000. A 90-day drinking water toxicity study in rats of the environmental contaminant ammonium perchlorate. *Toxicological Sciences* 57(1): 61-74.

U.S. EPA. 2000. Proposed Rule: National Primary Drinking Water Regulations; Arsenic and Clarifications to Compliance and New Source Contaminants Monitoring. *Federal Register* 65(121): 38887-38983.

Definitions

bioaccumulation: the buildup of toxic chemicals in living things

carcinogen: a substance that produces or contributes to cancer.

chronic: long-term

environmental epidemiology: the study of the effect on human health of physical, biologic, and chemical factors in the external environment, broadly conceived

immunotoxin: a chemical or agent that is toxic to the immune system.

LD_{50}: in a toxicity study, the lethal dose of the chemical to 50% of the test organisms.

morbidity: illness

mortality: the death rate in a population

neurotoxin: a chemical or agent that can damage or destroy nerve tissue

strain: a group of organisms that form part of a race or variety and are distinguished from other related organisms by some distinguishing feature

teratogen: a chemical or agent that can cause damage during embryonic growth or development

toxicology: the science that deals with poisons and their effects and antidotes.

toxin: any poison formed by a living organism.

Endocrine-Disrupting Compounds in Your Water?

Shane Synder

A number of compounds released into the environment by human activities can modulate endogenous hormone activities. They have been termed endocrine-disrupting compounds (EDCs) [1]. Environmental EDCs have been defined as exogenous agents which interfere with the "synthesis, secretion, transport, binding, action, or elimination of natural hormones in the body that are responsible for the maintenance of homeostasis, reproduction, development, and/or behavior" [2]. It has been hypothesized that such compounds may elicit a variety of adverse effects in both humans and wildlife, including promotion of hormone-dependent cancers, reproductive tract disorders, and reduction in reproductive fitness [2]. The most well documented types of EDCs are estrogenic compounds that are able to mimic or block the effects of natural estrogen. These effects were observed in reproductive abnormalities in aquatic organisms such as fish and certain reptiles [3-5].

Lake Mead, Nevada, is the largest reservoir on the Colorado River by volume (approximately 36.7 x 10^9 m^3) and covers approximately 593 km^2 of surface area. Lake Mead was formed in 1935 with the impoundment of the Colorado River by the Hoover Dam. The Colorado River provides approximately 97% of Lake Mead's water. The remaining water originates from the Virgin and Muddy Rivers (ª 1.5%), which discharge into the Overton Arm, and the Las Vegas Wash (ª 1.5%), which discharges into the Boulder Basin. More than 22 million people depend on water from Lake Mead and the lower Colorado River for domestic and agricultural usage [6]. Tertiary-treated wastewater, storm water, urban runoff, and shallow groundwater enter the Las Vegas Bay of the Boulder Basin of Lake Mead via the Las Vegas Wash. The Las Vegas Wash is a 19-km channel that drains the entire 4100 km^2 Las Vegas Valley hydrographic basin. Approximately 5.8 x 10^8 L of water per day (excluding storm water) flow through the Las Vegas Wash.

In 1996, feral carp from the Las Vegas Wash and Las Vegas Bay were reported to have significantly different plasma sex steroid and vitellogenin (an egg yolk protein) levels than carp collected from a reference site in Callville Bay [7]. These types of endocrine disruptive effects have been associated with estrogenic substances found in wastewater effluents [4, 8-11]. Concentrations of some synthetic organic chemicals were found to be greater in water, sediment, and fish tissues from the Las Vegas Wash and Las Vegas Bay when compared with similar samples from the Callville Bay reference site [7]. Further studies revealed that natural estrogen and synthetic estrogen (from oral birth control medications) were entering the Las Vegas Bay in trace quantities [12, 13]. Furthermore, genetically engineered cells that detect estrogenic substances determined that the natural and synthetic estrogens were the most potent estrogenic compounds in these waters [14]. However, extensive screening of the drinking water in Southern Nevada did not detect any of these estrogenic compounds [15]. In 1998, it was determined that a variety of pharmaceuticals and personal care products were also present in the Las Vegas Bay [16]. Recently it has been shown that these compounds are present ubiquitously in all wastewater effluents [17]. The quantities of these substances are extremely small and can only be detected by modern analytical instrumentation. Research continues to determine whether these compounds have toxicological relevance in Lake Mead.

References

1. Kendall, R.J., et al., *Principles and Processes for Evaluating Endocrine Disruptors in Wildlife*. 1998, Pensacola, FL: SETAC Press. 491.
2. *Special report on environmental endocrine disruption: an effects assessment and analysis.* Office of Research and Development. Washington, D.C. February 1997. EPA/630/R-96/012, 1997.
3. Purdom, C.E., et al., *Estrogenic effects of effluents from sewage treatment works.* Chemistry and Ecology, 1994. 8: p. 275-285.
4. Jobling, S., et al., *Widespread Sexual Disruption in Wild Fish.* Environmental Science & Technology, 1998. 32: p. 2498-2506.
5. van Aerle, R., et al., *Sexual disruption in a second species of wild cyprinid fish (the gudgeon,* Gobio gobio*) in United Kingdon freshwaters.* Environmental Toxicology and Chemistry, 2001. 20(12): p. 2841-2847.
6. LaBounty, J.F. and M.J. Horn, *The Influence of Drainage from the Las Vegas Valley on the Limnology of Boulder Basin, Lake Mead, Arizona-Nevada.* Journal of Lake and Reservoir Management, 1997. 13(2): p. 95-108.
7. Bevans, H.E., et al., *Synthetic Organic Compounds and Carp Endocrinology and Histology in Las Vegas Wash and Las Vegas and Callville Bays of Lake Mead, Nevada, 1992 and 1995.* Water-Resources Investigations Report 96-4266, 1996.
8. Talmage, S.S., *Environmental and Human Safety of Major Surfactants: Alcohol Ethoxylates and Alkylphenol Ethoxylates*. 1994, Boca Raton: Lewis Publishers. 374.
9. Routledge, E.J., et al., *Identification of estrogenic chemicals in STW effluent. 2. In vivo responses in trout and roach.* Environmental Toxicology and Chemistry, 1998. 32(11): p. 1559-1565.
10. Giesy, J.P. and E.M. Snyder, *Xenobiotic modulation of endocrine function in fishes*, in *Principles and Processes for Evaluating Endocrine Disruption*. 1998, SETAC Press. p. 155-237.
11. Folmar, L.C., et al., *Vitellogenin Induction and Reduced Serum Testosterone Concentrations in Feral Male Carp* (Cyprinus carpio) *Captured Near a Major Metropolitan Sewage Treatment Plant.* Environmental Health Perspectives, 1996. 104(10): p. 1096-1101.
12. Snyder, S.A., et al., *analytical methods for detection of selected estrogenic compounds in aqueous mixtures.* Environmental Science & Technology, 1999. 33(16): p. 2814-2820.
13. Snyder, S.A., et al., *Instrumental and Bioanalytical Measures of Endocrine Disruptors in Water*, in *Analysis of Environmental Endocrine Disruptors*, L.H. Keith, T.L. Jones-Lepp, and L.L. Needham, Editors. 2000, American Chemical Society: Washington, DC. p. 73-95.
14. Snyder, S.A., et al., *Identification and quantification of estrogen receptor agonists in wastewater effluents.* Environmental Science & Technology, 2001. 35(18): p. 3620-3625.
15. Roefer, P., et al., *Endocrine-disrupting chemicals in a source water.* Journal American Water Works Association, 2000. 92(8): p. 52-58.
16. Snyder, S.A., et al., *Pharmaceuticals and personal care products in the waters of Lake Mead, Nevada*, in *Pharmaceuticals and Personal Care Products in the Environment: Scientific and Regulatory Issues*, C.G. Daughton and T.L. Jones-Lepp, Editors. 2001, American Chemical Society: Washington, D.C. p. 116-140.
17. Daughton, C.G. and T.A. Ternes, *Pharmaceuticals and Personal Care Products in the Environment: Agents of Subtle Change?* Environmental Health Perspectives, 1999. 107(6): p. 907-938.

CHAPTER 7

Air Pollution in Nevada

Randy Smith

This chapter will explore the topic of air pollution in Nevada mostly from the perspective of those who live in the two major metropolitan areas of the state, Las Vegas Valley (Henderson/North Las Vegas/Las Vegas) and Truckee Meadows (Sparks/Reno).

As you begin to read this chapter on air pollution, please pause for a moment and consider the question "What is pollution?" Write down your definition for pollution. Now that you have thought about it, does your definition only consider human caused situations, or is your definition more broad based, including nonhuman caused situations such as gasses and ash spewing from a volcano? Whatever definition of pollution you consider yours, information about air pollution tends to focus on human caused situations because it is these situations we may have control over and may be able to improve.

Take a deep breath and consider whether the air you just inhaled is unpolluted or polluted. What do you think? If you are reading this outdoors, it is probable you are breathing less polluted air than if you were outdoors reading this during the years of the 1970s, '80s, or '90s. Generally speaking, air quality is better now than it was during those years. However, this is not to say that the air may not be as clean as you might like. In fact, it may be considered unhealthy by regulatory agencies such as the U. S. Environmental Protection Agency (USEPA), Clark County Health District, or Washoe County Health District. (Take a look across the Las Vegas valley if you are in the Las Vegas area, or look towards Mt. Rose if you are in the Reno area. If you can see the air, and not the surrounding mountains, then there is a good chance that the air is not at the level of quality you would like.) If you are inside you might think that the answer to the question asked above is obvious: you are breathing in unpolluted air. To the contrary, if you are reading this in a new building, in a casino, or in an establishment that permits smoking, then there is an excellent probability you are breathing in polluted air.

Why are we even concerned with the condition and quality of our air? The names of the regulatory agencies mentioned above should provide a hint as to why we as a society are concerned about our air quality. The relative quality of the air we breathe affects our individual and collective health as well as the environment. For example, throughout our nation hospital admissions rise on days of poor air quality when compared to admission rates when the air quality is good. Air pollution can affect our environment in very surprising ways. A recent report (January 2002) suggests that pollution sources in California, Nevada, and Idaho may be playing a role in the survival of Big Horn Sheep lambs in the Wind River Mountains of Wyoming.

You have read in your environmental science textbook that despite the fact that air quality conditions are generally better than previous years, there are many air pollution problems in our country. The same is true for Nevada. Air quality conditions and situations in Nevada will now be discussed, with the focus on the metropolitan areas of Las Vegas and Reno.

Las Vegas Valley

The Las Vegas metropolitan area is in the county of Clark. In Nevada, counties are responsible for monitoring and regulating air quality conditions; historically this has been done by the Clark County Health District. Recently (July 2001) this responsibility has been given to the Clark County Commissioners. In Clark County, the major air pollution problems of the five national criteria pollutants (reread your primary text if you have forgotten what these are) concern carbon monoxide (CO), particulate matter (PM 10 and PM 25), and ozone (O_3).

Carbon monoxide is an odorless, colorless, tasteless, gas derived from the incomplete combustion of fuel. What happens is that the more efficiently the fuel is burned, the less CO is released; the less efficiently the fuel is burned, the more CO is released. The fuel may be natural gas, gasoline, diesel, coal, wood, and others. The greatest contributor of CO in Clark County is the automobile (approximately 74%). Another important contributor is the wood burning fireplace. Automobiles tend to burn fuel most efficiently in the speed range of 25-50 miles per hour (mph). What percent of your driving time is spent in this range? Under and over those speeds the fuel is burned less efficiently therefore more CO is emitted from the tailpipe. When driving on highways through town (US 95, US 93, I-15, I-215, I-515) most are driving faster than 55 mph. When you are stopped at a traffic light or are in line at a drive-through window you are going 0 mph and not burning fuel efficiently. To give you a feeling for the relative importance automobiles have on CO emissions, let us examine some statistics. According to the Clark County Health District, for the year 2000 an estimated 24.9 million miles were driven in the Las Vegas valley *everyday* resulting in 620,400 pounds of CO being released from automobiles alone *everyday*. To put into perspective how much CO vehicles release relative to other sources, TIMET, located in Henderson, is the largest single CO releaser in the valley. TIMET releases 105,000 pounds per *year*. Automobiles release 620,400 pounds everyday; TIMET releases 105,000 per year. That is certainly something to ponder.

We are concerned about CO because at relatively low concentrations, it is toxic. The U.S. EPA has set danger levels at 9 parts per million (ppm) average over an 8 hour period. CO exerts a double whammy on our bodies. Our bodies deliver oxygen to their cells via the blood by a blood protein called hemoglobin, inside the red blood cells. CO has a 210 times greater binding affinity for hemoglobin than does oxygen. What this means is that when molecules of CO and oxygen are in the presence of hemoglobin, the CO molecule is more likely to bind to the hemoglobin than is the oxygen molecule. (Imagine that hemoglobin is a piece of iron and CO and oxygen are magnets. CO is the much stronger magnet.) So, as you might surmise, if the concentration of CO is relatively high, it binds to the hemoglobin. Enough oxygen will not bind to the hemoglobin, thus depriving your body cells of needed oxygen. If your cells don't have adequate oxygen, they cannot derive enough energy from the foods you have eaten to survive. That is the first whammy. The second whammy is that it reduces the binding affinity of the oxygen molecules that have bound to the hemoglobin. What this means is that the oxygen will likely become uncoupled from the hemoglobin before it reaches its appropriate destination. Fortunately, outdoor CO concentrations typically do not get high enough for death to result (Indoors is another story. Read in your main textbook about the dire consequences that can happen if CO concentrations get too high indoors.). However, drowsiness, severe headaches, dizziness, nausea, slower reflexes, and slower thinking capabilities are symptoms that can occur.

Clark County is considered to be in non-attainment for CO, meaning that its CO concentrations are, or at one point were, too high per stated regulatory levels (9 ppm per cubic meter per 8 hour period). This designation of non-attainment comes primarily from three exceedences at the East Charleston monitoring site during 1996. Historical data suggest the evening hours from November through January are when the highest CO concentrations are most likely. Think of some reasons why these times would produce the greatest CO concentrations.

Clark County is also considered to be in non-attainment for particulate matter. Particulate matter is dust. Particulate matter comes in different sizes. As you drive on a dry dirt road dust will waft in the air. Some of the dust particles are large enough to fall quickly back to the ground. Others stay afloat in the air longer but if inhaled are filtered by hairs and mucous. The smallest particles, **P**articulate **M**atter of **10** microns and smaller (**PM 10** — this is the category for which Clark County is in non-attainment.) and **P**articulate **M**atter of

2.5 microns and smaller (**PM 2.5**) can be inhaled and can get by your natural filter system. (What is your natural filter system?) PM 10 sized particles may be expelled from your body over time; however, those smaller than that, including the PM 2.5 particles, can be drawn deep into the lungs and become lodged there. This is the most dangerous type of particulate matter. Inhalation of particulate matter can provide a double whammy situation as well. Firstly, particulate matter lodging in the lungs creates its own special set of problems such as tissue damage. Secondly, particulate matter tends to be charged, resulting in some charged pollutants adhering to it. In this situation the pollutant molecules piggyback on the particulate matter. When particulates are inhaled, they end up delivering the pollutant to the lungs. Whammy one, the particulate gets lodged in the lungs. Whammy two, the particulate may be carrying a pollutant. Chronic exposure to particulate matter can lead to adverse respiratory conditions such as emphysema and bronchitis.

Before you continue with your reading, pause for a minute or two and write down what you would consider the major and lesser contributors to particulate matter pollution in the Las Vegas valley. Earlier it was stated that if you could see the air, then the air was probably not as clean as you would like it. When present, much of the haze over the Las Vegas Valley is particulate matter. This is because small particulates scatter light. The greatest contributors to the particulate matter in the Las Vegas Valley are construction activity and windblown dust from the surrounding areas. Other lesser but still important contributors are paved and unpaved roads, smoke from wood burning, ammonia salts, and gasoline and diesel vehicle emissions. (How did you do in identifying the contributors to particulate matter pollution?)

Ozone (O_3) is considered a secondary air pollutant because it is formed from sun stimulated chemical reactions between primary pollutants. Remember that primary pollutants are pollutants that come right out of the tailpipe or smokestack. The primary pollutants involved in making O_3 are hydrocarbons (VOCs) and nitrogen oxides. Ozone concentrations tend to be worse on winter **inversion** days and hot summer afternoons. (Any of those in Clark county?) Clark County is in attainment for O_3.

A pollutant that causes problems for many in the Las Vegas Valley but is not considered a criteria pollutant is pollen. For allergy sufferers, pollen may cause much more acute discomfort during certain times of the year than criteria pollutants. In the 1950s, '60s, and '70s, physicians often suggested their asthma and allergy patients move to communities in the southwest desert to find relief from their medical conditions. As people moved to the southwest desert communities, they planted trees that were prodigious pollen producers such as fruitless mulberry trees and European olive trees. The planting of these two in the Las Vegas Valley has been banned since 1992. Even with such regulations, pollen can reach extremely high levels. High tree pollen counts may be present from late January through early June. Grass pollen is at its highest from March through October. (See the Clark County Health District website for more information about pollen and associated allergies.)

Truckee Meadows

The cities of Reno and Sparks are in Washoe County, and therefore the Washoe County Health District monitors its air quality. Important pollutants in Washoe County are particulate matter, CO, and O_3.

Reno/Sparks (Truckee Meadows) are situated in a basin with mountains on three sides. Particulate and carbon monoxide air pollution problems in Washoe County are worst during the fall and winter months (primarily November and December). This is because of conditions that result in frequent inversions. (Quick. Jot down what an inversion is.) Reduced air quality caused by O_3 is most frequent in the summer.

Particulate matter air pollution is of the greatest concern in the Truckee Meadows. Remind yourself of the problems associated with particulate matter. The Truckee Meadows is designated as a non-attainment area for PM10 by the USEPA. Road dust from paved streets is the major contributor to particulate matter concentrations on an annual basis. The second major contributor is construction activities, followed by residential wood combustion. However, the major contributors of particulate matter in the winter months when the problem is the worst, are residential wood combustion first, and paved roads, second.

To combat particulate matter pollution in the winter months when this type of pollution is most serious, Washoe County has implemented a program that restricts wood burning during certain air quality conditions. It also educates its citizens about their role in reducing particulate matter pollution in the Truckee Meadows. The program is called Green, Yellow, Red. The colors correspond to the colors of a traffic signal. On Green designated days, wood burning can take place. On days designated as Yellow, the public is asked to voluntarily refrain from burning wood and to cut back on car trips. Red days signify no burning. Fines or warnings are issued to offenders. The program has been quite successful and has played an important role in reducing the number of designated unhealthy days in the Truckee Meadows.

The concern over CO has been discussed earlier and in your primary textbook. The greatest contributors of winter CO in Reno/Sparks are automobiles and wood burning stoves. Eighty four percent of the problem is attributed to motor vehicles (including both on road and non-road vehicles), much of the remaining to wood burning stoves. In the Truckee Meadows, 180,681 pounds of CO are released everyday from motor vehicles traveling 8.3 million miles per day. The Reno/Sparks urban area is considered to be in non-attainment for CO. The last exceedence for this pollutant was December 13, 1991. Remember an area is considered non-attainment if it has exceeded the National Air Quality Standards (NAAQS) for a particular pollutant. Since the Reno/Sparks area has had an exceedence for CO, it is in non-attainment for this pollutant and is referred to as the Truckee Meadows non-attainment area.

Ozone concentrations reach their highest in the Truckee Meadows during the summer months of June, July, and August. Remember that ozone is a secondary pollutant derived primarily from sun powered chemical reactions involving volatile organic compounds (VOCs) and nitrogen oxides (NO_x). The major contributors for both of these primary air pollutants are motor vehicles (both on-road and non-road). The Truckee Meadows is designated as being in marginal non-attainment for Ozone. Low level concentrations of ozone causes eye irritation, and chronic exposure may lead to respiratory problems. Ozone is a major contributor to reducing the life of tires and other rubber products by causing hardening and cracking.

Indoor Air Pollution

Mentioned earlier was the idea that people often assume that air quality indoors is better than outdoors. Contributing to this assumption is the fact that local air pollution regulatory agencies suggest people remain indoors during days when the outside air is deemed unhealthy. However, under certain conditions indoor air pollution may be much more severe than the outside conditions. This is important because on average, people spend about 90% of time indoors while elderly and housebound people spend nearly all their time indoors. Typically indoor air pollutants are different pollutants from outdoor air pollutants. Indoor air pollutants include such things as formaldehyde, constituents of cigarette smoke, glues, asbestos, air fresheners, paints, lead, radon, and biologicals such as molds and pet dander. All of these mentioned can be in concentrations high enough to cause health problems.

You can learn about many of the above mentioned indoor air pollutants by reading your main Environmental Science textbook. This segment of the chapter will discuss indoor air pollutants that are likely to be more prevalent in the Las Vegas Valley and Truckee Meadows than in other locales in the United States: cigarette smoke, glues, paints, and formaldehyde.

Let us begin our discussion with cigarette smoke. If you frequent casinos for entertainment or you are an employee in a casino, you are aware that cigarette smoke is a common indoor pollutant in both the Las Vegas Valley and Truckee Meadows. Cigarette smoke contains over 4,000 different chemicals. Many are strong irritants. Two hundred are known poisons and 43 have been demonstrated to cause cancer in humans and other animals. Secondhand smoke, also known as Environmental Tobacco Smoke (ETS) is divided into two types; mainstream smoke and sidestream smoke. Mainstream smoke is the smoke that is inhaled and subsequently exhaled by the smoker. Sidestream smoke goes directly into the air from the burning tobacco source. It is the sidestream smoke that is most prevalent in room air and carries high concentrations of dangerous chemicals. Why do you suspect sidestream smoke is more dangerous than mainstream smoke?

Secondhand smoke has been implicated in increasing the risk of lung and other cancer for those who have long time exposure. In fact, the United

States Environmental Protection Agency (USEPA) classifies it as a known cause of lung cancer in humans. The American Lung Association estimates that ETS is the causation of approximately 3,000 lung cancer deaths and 35,000 – 50,000 heart disease deaths in nonsmokers each year. Nevadans are more likely to encounter secondhand smoke because, according to the Centers for Disease Control, the state has some of the weakest tobacco laws in the US. Smoking is allowed in most places the public frequents. Some exceptions are government buildings, school buses, and public elevators.

Children are the most vulnerable members of our population for exposure to ETS. Nevada, with about 32% of its adult population being smokers, has one of the highest smoking rates in the United States. This suggests that many children in Nevada may be exposed to ETS. Children exposed to ETS are more likely to suffer from reduced lung function, respiratory tract infections, ear infections, sudden infant death syndrome (SIDS), and asthma problems: increased rates, increased episodes, and more pronounced symptoms. The higher risk for the above mentioned conditions leads to 5.4 million excess cases of childhood disease. Treatment of these health problems requires 8 percent of all pediatric medical spending. Approximately 6,200 excess childhood deaths are also attributed to chronic ETS exposure.

The State of Nevada has one of the highest rates of human population growth in the United States. Accompanying this growth is the high rate of new construction projects you witness in the Las Vegas Valley and Truckee Meadows. Population growth and new construction translates into many in Nevada living in new homes, working, or going to school in buildings that have been newly constructed. A number of chemicals volatilize (look up this word in your dictionary) from construction materials and new furnishings. Many household furnishings emit chemicals throughout the time they are in our homes but most profoundly when they are new. Such items as couches, chairs, curtains, carpets, and beds volatilize the chemical known as formaldehyde. Formaldehyde is used as an adhesive, bonding agent, and preservative in such products as furniture, carpets, and items made of particle board. Formaldehyde can cause eye and respiratory irritation, headaches, rashes, and a variety of other symptoms. Many builders and furniture manufacturers are now using low formaldehyde products. To be safe and informed, ask about the presence of formaldehyde in products being used to build your new house or new furnishings for your house.

What Can You Do?

Outdoor Air Pollutants

After reading the prelude to this section you may be afraid to take a breath of air. We all know this is not possible, so let us look at ways you can be part of the solution. Before you read further, write down on a separate sheet of paper your good ideas about what you can do to play a positive role in improving the air pollution problem, both outdoor and indoor. See how your list compares with the items discussed below.

When reading about outdoor air pollution, it should have struck you that the use of the automobile is responsible for much of the problems concerning CO, ozone, and particulate matter. If attention is focused on the use of the automobile, then improvements in this area could translate to improvements in the reduction of CO, ozone, and particulate matter concentrations. What you can do is not use your automobile as much as you do currently. Implementation for reduction can begin in the planning of your daily trips. Make your trips as efficient as possible. If you are in an area of town that has stores where you need to shop, shop at the stores while you are in the area instead of making separate trips. Participate in carpooling for going to school or work. Do not use drive-through windows, especially when a number of cars are in line. Remember, while you are sitting there idling, your vehicle is getting zero miles to the gallon. Besides, often you can complete your transaction more quickly and save time by going inside.

Buy an automobile that gets high miles per gallon. Better mileage translates into fewer emissions. Along the same line of thought, keep your tires inflated to recommended pressures. When this is done, less friction occurs and your gas mileage improves. Doing regular maintenance on your vehicle can also result in improved gas mileage.

Write your elected representatives and senators to advocate improved gas mileage standards for vehicles, the end to emission exemptions for diesel vehicles, funding for alternatives such as ethanol and fuel cell technology, and incentives for increasing carpooling and mass rapid transit.

Report cars releasing smoky exhaust. In Las Vegas, dial (702) 642-SMOG and in Reno/Sparks dial (775) 686-SMOG. When you call be prepared to provide the license plate number of the car, and time and location you saw the offender. For the first reporting, a letter will be sent to the owner of the vehicle. Upon a second reporting, an investigator will be sent to confirm there is a problem. If a real problem exists, the owner has 60 days to remedy the problem or the vehicle registration will be revoked. This program is present because it is estimated that a small percentage of the cars are responsible for a large percentage of the problem. Don't be ashamed to report these vehicles; what comes out of their tailpipe goes into your lungs.

The above suggestions focus on the automobile. Other things you can do to become part of the solution is refrain from using your fireplace or wood burning stove on cool, calm evenings when conditions are ripe for an inversion. As mentioned previously, the Truckee Meadows implemented the Green, Yellow, Red program to inform its citizens when it is permissible or not permissible to burn. Those living in the Las Vegas Valley will have to educate themselves about how climatic conditions correspond to unfavorable burn days.

Indoor Air Pollutants

Earlier you learned that cigarette smoke is a major indoor air pollutant in Nevada. You can reduce your exposure by not frequenting buildings where secondhand tobacco smoke is prevalent. Vote with your pocketbook; support businesses that do not allow smoking in their establishments. Ask businesses that do allow smoking to reduce the amount of area where smoking is permitted. If you work in an environment where environmental tobacco smoke is present, you are in a tough situation. You can continue to work in such an environment, thus continuing to be exposed to ETS. You can quit and find employment in a workplace with little or no ETS, or you can try to work with your employer to either change their policy or station you in an area that may not be quite so smoky. These are difficult choices to make because they may affect your long-term health and your ability to provide for you and your family.

It may be easier to be more proactive in maintaining healthy air quality in your place of residence than at your workplace. At home you can buy products that do not contain formaldehyde or work with your builder to use products containing low concentrations of formaldehyde or none at all. You can make sure the air in your home is circulated regularly. If a spouse or roommate smokes, have them do so outside. Be sure to ventilate your house when using products containing VOCs. These products include paints, household cleaners, pesticides, hobby products, and glues. Read your main textbook for further information concerning pet dander, radon, asbestos, biologicals, and carbon monoxide.

In this chapter, air quality for the Las Vegas valley and Truckee Meadows was discussed. Pollutants of concern in these areas are CO, particulate matter, and ozone. Two important indoor air pollutants were also reviewed. Measures for improvement were discussed for both outdoor and indoor pollutants.

References

Air Quality Management Division. 2001. Washoe County, Nevada; Air quality trends 1989-2000. Washoe County District Health Department.

American Lung Association. 2002. http://www.lungusa.org

Associated Press. 2001. Nevada officials considering vote on smoking restrictions. *Las Vegas Sun*, November 9, 2001.

Carola, R., J.P. Harley, and C. R. Noback. 1992. *Human Anatomy & Physiology*, second edition. McGraw-Hill, Inc. New York. 977 pp.

Center for Disease Control and Prevention. 2002. http://www.cdc.gov

Clark County Health District. 2001. http://www.cchd.org

Cunningham, W.P. and B. W. Saigo. 2001. *Environmental Science, A Global Concern*, sixth edition. McGraw-Hill, Inc. New York. 646 pp.

Ellyson, M. 1993. Personal communication. Clark County Health District.

Goodrich A. and L. O'Brien. 1996. Washoe County, Nevada: 1995 emission inventory for particulate matter. Washoe County District Health Department.

Hill, J.W. and D.K. Kolb. 1995. *Chemistry for Changing Times*: Seventh edition. Prentice-Hall, Inc. New Jersey. 720 pp.

Ling, Y. and L. O'Brien. 2001. Washoe County, Nevada. Ozone Non-Attainment Area: 1999 periodic emissions inventory of ozone precursors. Washoe County District Health Department.

Ling, Y. and L. O'Brien. 2001. Truckee Meadows Non-Attainment Area: 1999 emissions inventory for carbon monoxide. Washoe County District Health Department.

Manning, M. 2001. Smog sends many residents to ERs. *Las Vegas Sun*, October 2, 2001.

Montana State University. 2002. http://www.Montana.edu/wwwcxair

Polakovic, G. 2001. Deaths of the Little Bighorns: A mysterious illness is weakening lambs in the Rockies, with many falling prey to predators. Researchers say pollution may be the cause. *Los Angeles Times*, August 29, 2001.

Smolinski, R. 2002. Personal communication. Clark County Health District.

United States Environmental Protection Agency. 2002. http://www.epa.gov/ebtpages/air.html

Washoe County Health District website. 2001. http://www.co.washoe.nv.us/Health

Washoe County District Health Department. 2000/2001 Annual Report.

Watson, J. 1997. Personal communication. Desert Research Institute, Reno, Nevada.

Chapter 8

Nevada Geology and Earth Resources

Ed Eschner

Nature Makes Chemical Compounds Too

Often we read or hear about the chemical hazards humans have created along with the technological revolution. News of toxic waste dumps, repositories, spills, and air emissions fill our newspapers and textbooks. But long before humans appeared, Mother Nature was busy making the **elements** and mixing them to form **chemical compounds**. Some of these elements and compounds have qualities that are hazardous to life forms, others are essential to life forms, and still others are just pleasing to look at.

When we find elements and compounds that are naturally occurring, **inorganic crystalline** solids, that have a definite internal ordered arrangement of **atoms**, we call them **minerals**. Because minerals have ordered internal structural arrangement of atoms, they may form beautiful geometric shapes called **crystals**. Crystals form either by **precipitating** from **magma** or from water.

Minerals may be formed of single elements such as gold (Au) or silver (Ag) or copper (Cu). However, because atoms of different elements sometimes join, or **bond**, minerals may be made from compounds of two or more elements. Minerals are classified based on chemical composition.

Three Main Rock Types

Minerals form the building blocks for **rocks**, and rocks may be built out of one or more types of minerals. Geologists have categorized three main rock types based upon how the rock formed and the type of environment in which the rock formed.

Igneous Rocks

Igneous rocks form from hot, gooey chemical cookie dough called magma. Magma is generated deep beneath the earth's surface and it rises and intrudes into the surrounding "country rock," like a hot air balloon rises into the sky. Rising magma often takes the shape of a hot air balloon and we call it a **Pluton**, after the god of the underworld, Pluto.

Intrusive igneous rock forms if the magma cools off before reaching the surface. Beneath the blankets of earth **crust**, the magma slowly precipitates minerals that grow large enough to see without the aid of microscopes, although a good magnifying hand lens is often useful. The intrusive igneous rock is called a "**Plutonic**" rock. **Granite** is the name given to a light colored, silicate rich plutonic rock that forms this way. Granite reminds me of chocolate chip cookies. Humans find that granite is useful for making statues and countertops. Also, I enjoy shopping at the Galleria Mall in Henderson because of the beautiful granite floor tiles! There are very few natural outcroppings of granite in Southern Nevada, but a large exposure of granite does show in Boulder City. It is named the "Boulder City Pluton." Boulder City locals call the outcrop "B-Hill."

The Sierra Nevada Range, which borders Nevada to the west, is a massive north-south trending exposure of many plutons. Such large expanses of granite plutons (larger than 100 km^2) are call batholiths. Gold (Au) and silver (Ag) may form vein deposits in plutonic rock. The 49ers were searching for the mother vein, or mother lode, in the Sierra Nevada.

Granite is a plutonic igneous rock that forms from **siliceous** magma that cools slowly and crystallizes beneath the surface of the earth. But what if that magma reaches the surface and erupts, or extrudes out on to the surface? Such eruptions or extrusions are often very fiery and may be explosive (e.g., the 1980 eruption of Mount St. Helen's in the Cascade Range). Extruded igneous rock tends to crystallize quickly at or near the cool earth's surface; thus the mineral crystals do not have the luxury to grow very large. Often we will need the aid of a microscope to see individual minerals in a sample of extrusive igneous rock. **Extrusive igneous** rocks are also called **volcanic** after the god of fire, Vulcan. Never let anyone say that geologists are not religious.

So, if a siliceous magma like that which granite forms from does not solidify slowly beneath the surface, but instead reaches the surface and extrudes, the minerals quickly precipitate small crystals. They then form the extrusive igneous rock known as **rhyolite**.

Rhyolite often forms from explosive eruptions because the silica rich magma that it is made from is thick and sticky or viscous and gaseous. When rhyolite type eruptions occur, it is often like shaking a champagne bottle—then popping the cork. Sometimes important minerals like the native element gold come out in the explosive mess. Sometimes rhyolite erupts violently, sending ash spewing up where it is spread laterally by the winds. Hot ash that comes down and compresses and welds into a hard form is called **tuff.** Much of southeastern Nevada is composed of thick sheets of rhyolite, especially visible around Caliente, Kershaw-Ryan State Park, Pioche, Beaver Dam State Park, and Eagle Valley reservoir, and stretching westward towards Beatty. In fact, there is a ghost town near Beatty called Rhyolite, named after the gold-bearing extrusive igneous rock.

The town of Rhyolite was once a bustling mining town with quite a large population. Many permanent structures were built including grocery stores, banks, brothels, and a jail. Only partial structures remain, except the well-preserved train station and famous bottle house. The gold proved too difficult to extract without the modern method of **heap and cyanide leach**. Until its closure in the mid 1990s, the heap leach method was used at the Bullfrog Mine, located adjacent to the old town site of Rhyolite. Other volcanic features in the area include the Lathrop Wells cinder cone, a cone shaped deposit of volcanic ash, and Ubehebe Crater in Death Valley National Park, a bowl shaped depression formed by the explosion that resulted when hot rising magma flashed groundwater into steam. Mammoth Mountain, in the southern Sierra Nevada of eastern California, is a large volcano and ski resort with an explosive history. It is interesting that Yucca Mountain, the nation's proposed high-level nuclear waste repository, is a volcanic mountain.

We have learned about plutonic or intrusive, and volcanic or extrusive igneous rocks. You should be able to recognize the rock name for the examples given for rocks forming from a silica rich melt. However, please be aware that as the silica composition of the melt changes so does the resultant rock name.

Sedimentary Rocks

Rocks exposed at the earth's surface are broken down, or **weathered**, both mechanically and chemically. **Mechanical weathering** breaks down rock by physically cracking the rock in some way. For example: perhaps you have left a bottle of your favorite beverage in the freezer a little too long—the liquid inside the container expands when it crystallizes into ice and the container breaks. Rocks do the same thing as the bottle. If water penetrates cracks in the rock and if temperatures fall below freezing, the water expands and makes the crack in the rock bigger. The net result of weathering is that it takes large rocks and breaks them down into many smaller rocks called sediment. When this happens the surface area of the rock is increased and the rock is more susceptible to chemical weathering. Chemical weathering attacks the chemical composition of the minerals. This may result in some of the minerals that make the rock being dissolved and carried away in solution, by surface or ground water.

Weathering and gravity attack outcrops of rock that are exposed in areas of greater elevation. The rock material is broken down and transported by conveyor belts such as streams, mudslides, glaciers, and winds—towards areas of lower elevation such as basins, lakes, or ultimately the ocean. The further the sediment is transported from the source outcrop, the smaller and more rounded the individual grains of material tend to become. Thus, we find a sorting of material as we move away from the source outcrops. Typically, the grains, or **clasts**, are large angular boulders closer to the source outcrop and the clasts become smaller and more rounded as we move toward the ocean. The boulders break down into cobbles, gravel, sand, silts, and clays when transported over long distances.

If the clasts become deposited and cemented or **lithified**, we can categorize the rock as a **clastic sedimentary** rock, depending on the clast size and degree of rounding caused by transportation. Clastic sedimentary rocks like **sandstone** remind me of sugar cookies or pecan sandies. Minerals that have been dissolved into water may later precipitate out and these form the nonclastic or chemical/biochemical sedimentary rocks.

Limestone is a common nonclastic sedimentary rock made from the mineral **calcite**. Limestone crops out widely in southern Nevada —especially north of Lake Mead and west toward the Spring Mountains. Frenchman and Sunrise Mountains as well as the Spring Mountains are largely limestone. In Nevada, gypsum, sand, gravel, and placer gold are important economic sedimentary deposits.

Metamorphic Rocks

Rocks that are exposed to heat and pressure by deep burial or that are located close to magma may alter into **metamorphic** rocks. Literally, "meta" means change and "morph" means form; so rocks that are exposed to heat and pressure inside the earth oven may become harder, and the minerals may recrystallize into a linear, parallel alignment called foliation. Exposed to increasing heat and pressure, the sedimentary rock **shale**, for example, will change to the metamorphic rock, **slate**. The change in form from the shale to slate is only slight. As the degree of metamorphism increases with increasing temperature and pressure, the hardened clay particles in the slate start to crystallize into shiny mica minerals. The resultant rock has a silvery sheen called a **phyllite**. Exposed to more metamorphism the mica crystals grow and become flaky and foliated. The rock is now called a piece of **schist**. As metamorphism reaches its peak, the light and dark minerals separate into parallel bands called gneissic foliation, and the rock is known as a nice piece of **gneiss**. Not all metamorphic rocks become foliated. Some are nonfoliated; for example, sandstone metamorphoses into **quartzite**, and limestone becomes **marble**.

When we are walking in the field and we find some nice gneiss, or a piece of schist lying around on the surface of the earth, we know that rock has made an incredible journey. We know this because the environment where heat and pressure is sufficient to form nice gneiss or pieces of schist is found deep beneath the surface of the earth. We may conclude that either erosion or tectonic activity may have removed the rocks that once covered the gneiss and schist so deeply. Foliated metamorphic rocks including pieces of schist and gneiss are exposed in the southern Virgin Mountains on the east side of Lake Mead, on Saddle Island near the Lake Mead Marina, and on the east side of Las Vegas Valley at an extremely significant outcrop called the Great Unconformity. But before we can really appreciate the significance of the Great Unconformity, let us consider geologic time.

Geologic Time Frame

Geology is a youthful science, only a couple hundred years old. Before the development of modern geologic thinking, the earth was believed to be only 6,000 years old. Today there are still folks that think the earth is only a few thousand years old. This belief is based on the work of Archbishop Ussher's in the 1600s in which he calculated from the Bible the generation times since Abraham. These people believe in a young earth.

However, early pioneers in geologic thinking, such as Sir Charles Lyell, Charles Darwin, and James Hutton, the "Father of Modern Geology," used the scientific method of thinking to work out the relative age of the earth based on spatial relationships observed in the field. These geologists developed the "relative geologic time" concept based on the following geologic laws:

1. original horizontality
2. lateral continuity
3. faunal succession
4. superposition
5. crosscutting relationships
6. uniformitarianism

Using these laws, the relative age of the earth can be divided into four main eras, based on the fossil evidence of life that is found in the rocks, especially sedimentary rocks. Eras may be further broken down into smaller units called periods.

Precambrian

The Precambrian Era, from approximately 4.5 billion years ago to about 0.6 billion years ago, is the oldest era of earth history, and rocks formed during this time contain scant evidence of life forms other than photosynthetic prokaryotes such as blue-green algae. It is possible that these early producers changed the earth's atmosphere, paving the way for life as we know it. As they consumed carbon in the form of CO_2, they polluted their environment with exhaust gasses, namely oxygen in the form of O_2.

Precambrian rocks form the ultimate basement rocks because they are the oldest. In Southern Nevada and the southwestern United States, most of the Precambrian rock is metamorphic gneiss and schist intruded by granite. The basement complex is beautifully exposed in the Grand Canyon area, where the Colorado River has eroded and cut like a knife through a layer cake. The rocks are named the Vishnu Schist and the Zoroaster Granite in the Grand Canyon, and they represent the core of a lofty mountain range that was eroded nearly flat by long exposure at the earth's surface. The rocks have been radiometrically age dated at about 1.7 billion years old. Eventually, approximately 600-550 million years ago (**mya**) the ocean transgressed onto the land from the west towards the east. Waves crashed on the beachfront property and that environment laid down a thin blanket of sandstone. Because roughly 1.2 billion years of earth history were lost to erosion, the contact between the Precambrian rocks and the overlying sandstone is called the Great Unconformity. With the rising tide in southern Nevada came another profound change. Around the globe the conditions were right, and oxygen had accumulated in sufficient concentration to allow an explosion of primitive life forms to bloom, thus heralding the end of the Precambrian Era and the beginning of the era of early life know as the Paleozoic, or early-life eras.

Paleozoic

Primitive life forms such as the horseshoe crab-like trilobite, clam-like bivalves and brachiopods, and solitary coral burst into bloom on the ocean bottom. Perhaps nutrient-rich hot springs known as seafloor vents provided a fertile nursery. As the Paleozoic era progressed, more and more sedimentary layers were deposited and more primitive life forms lived and died. Some of the bodies became buried and were fossilized. These fossilized remains allow sedimentary layers to be read like pages of a history book, and they form the foundation for relative geologic time. When geologists look closely at the fossils of the Paleozoic era, a progression of life forms can be seen, ranging from primitive invertebrates in the early Paleozoic to primitive vertebrates such as bony armored fish in the middle Paleozoic. Those primitive fish evolved into amphibians just before the end of the Paleozoic.

Throughout much of United States, younger sedimentary rocks were piled onto the basement of Precambrian rocks. The sedimentary layers were laid down, with each layer, or bed, recording the environment of the time. The oldest sedimentary layers were laid down closest to the basement, as can be seen so well at the Grand Canyon of Arizona. As one climbs from the bottom of the Grand Canyon to the top, one climbs through successively younger layers. The layers are named, from bottom to the top, the Tappeas Sandstone, the Bright Angel Shale, the Mauve Limestone, the Redwall Limestone, the Supai formation, the Hermit Shale, the Coconino Sandstone, the Toroweap formation, and finally at the rim, the Kaibab Limestone.

The Kaibab Limestone marks the retreat of the ocean at the end of the Paleozoic era. The closing of the Paleozoic is marked in the rock record by the disappearance of many aquatic and terrestrial species. The cause is poorly understood — perhaps catastrophic ecosystem collapse, perhaps a natural disaster such as volcanic eruption or meteor impact.

Mesozoic

The Mesozoic, or middle-life era is also known as "the age of dinosaurs." The dinosaurs, or "thunder lizards" (*dino-* = thunder; *-saur* = lizard) dominated the Mesozoic Era, which lasted from approximately 250 mya-65 mya. The Mesozoic Era is broken into three smaller time units called periods. The Mesozoic Era ends with the disappearance of many species from the rock record. This paucity of both aquatic and terrestrial species is widely attributed to a terrible meteor impact. This impact, along with associated volcanic eruptions, spewed dust and ash into the atmosphere, thus obscuring the sun and making the temperature plunge. Assuming the dinosaurs were cold-blooded like our modern reptiles and could not maintain their body temperature, they died off in the cold, and left mammals heir to the throne. It has been suggested that perhaps the dinosaurs did not all die out, but instead some may have evolved into organisms that we do not recognize as dinosaurs, such as birds.

The skeletal structure of modern birds closely resembles that of some dinosaurs. Most fossils contain only the hardest parts of the organism, namely the bones and teeth, so if dinosaurs were actually feathered, the chance that they would be preserved is rare. However, such a rare fossil was preserved in shale found in Germany. The fossil animal is called *Archaeopteryx* (*Archaeo-* = ancient; *pteryx-* = wings) and was once believed to be the first bird. The fossil clearly has feathers, wings, and a tail but the skeletal structure is that of a reptile, including teeth.

Another interesting Mesozoic reptile fossil is the ichthyosaur (*ichthyo-* = fish; *-saur* = lizard). The ichthyosaur was a large (up to 20 meters long) aquatic marine reptile with large eyes for hunting in the deep dark ocean. Well-preserved fossils show that the ichthyosaur gave birth to live young; a mammalian trait that may link this ancient reptile to whales. But the skeletal structure is that of a reptile. Such outstanding specimens of ichthyosaur are found in Nevada; the ichthyosaur is the state fossil of Nevada. Specimens may be observed in site at Berlin-Ichthyosaur State Park in central Nevada. The park is located approximately 150 km north of Tonopah. The town of Berlin is a ghost town left over from gold mining activities of the early 1900s.

In Berlin, several well-preserved buildings can be seen, including the stamp mill used to crush ore. It was the miners of Berlin who first discovered the ichthyosaur fossils. Charles Camp, a paleontologist from UC Berkeley, later excavated the site and found an assemblage of eight or nine large adult fossils. Later, approximately 30 more were found. Early interpretations of the rock deposits suggest that the ichthyosaurs were beached in shallow water, similar to the mass beaching of modern whales. An alternative explanation is they were hunting in a deep submarine canyon when they were buried in a subaqueous landslide called a turbidity current.

Regardless of how they died and regardless of weather they were reptiles or early aquatic mammals, the fossils preserved at Berlin-Ichthyosaur State Park are nothing less than miraculous. The fact that not just one, but many of these large creatures died and their relatively undisturbed bones were preserved intact and were lithified into fossils is incredible! But more than that, the Nevada State Fossil survived unbelievable tectonic forces.

During the end of the Mesozoic Era, the North American continent crashed head on with the Pacific Ocean tectonic plate. The results were disastrous. The North American plate yielded in the accident and the West Coast was wrinkled, folded, and faulted. The western portion was pushed eastward, the rocks broke along a series of north-south faults hundreds of miles. The rocks were "thrust" faulted as the western deep-water Paleozoic sediments were pushed up and over the younger Mesozoic sediments. Later, the older, higher sediments were eroded, and the ichthyosaur fossils were found in the window of the thrust sheet. In southern Nevada, the Mesozoic thrust fault is most clearly visible at the Red Rock Canyon Recreational Area and at the Valley of Fire State Park.

Ichthyosaurs are not found in southern Nevada because they are a Mesozoic marine species. Land emerged from the Paleozoic seas towards the closing of the era, after the deposition of the Kaibab, and the earliest Mesozoic rocks were deposited in a tidal flat type environment with fine grain sediments called the Moenkopi formation. Rivers poured out of highlands onto the tidal flat, depositing coarser grain sediments along with logs and branches from the mountains. The wood was quickly buried by siliceous volcanic ash and the logs became petrified by a process that replaces the molecules in the wood with silica, or rock. The result is a rock that looks like the wood it replaced. Sometimes great detail is preserved, including bark, knotholes, and tree rings. These petrified logs and wood pieces are found in the Chinle formation. The logs are exposed in the Chinle formation at the Valley of Fire State Park. However, the best examples were explored by John Muir in 1906 at what is now the Petrified Forest of Arizona. Dinosaur fossils are also found in the Chinle Formation.

After the episode that produced the Petrified Forest, the North American Continental Plate moved northward out of the tropics, from an area near the equator toward the desert belt. During the Jurassic Period of the Mesozoic Era, the region of North America became a desert complete with terrestrial sand dune fields similar to those found in the Sahara desert region today. Dry hot winds blew billions of grains of quartz sand into piles thousands of feet thick. As the sand dunes were piled up, they migrated across the land. The stacking of one dune surface over another resulted in the stunningly beautiful internal stratification known as cross-bedding.

Locally, this cross-bedded sand dune deposit is called the Aztec Sandstone. Fantastic outcrops of fiery red, iron stained and buff-white sandstone can best be seen at the Red Rock and Valley of Fire State Park of Southern Nevada. At both of those locations, Paleozoic marine sediments overlie the Mesozoic terrestrial sand dune deposits. This does not violate the law of superpositioning, because this inverted stratigraphy was caused by the crustal collision described earlier.

However, the sandstone was deposited across most of Southwestern North America. Geologists working adjacent to, and East of the Basin and Range province in the Colorado Plateau region, named the same formation the Navajo Sandstone. The Navajo Sandstone forms the spectacular cliffs at Zion National Park and at Snow Mountain State Park near Saint George, Utah.

Cenozoic

We are living in the Cenozoic Era. Cenozoic means "recent life" and the Cenozoic rocks begin where the dinosaurs disappear from the rock record, around 65 mya. Because geologists consider 65 million years to be "recent," you should never lend geologists money.

The Cenozoic brought many changes to the geologic landscape. The crust pulled apart following the Mesozoic continental collision with the Pacific plate. The thickened Mesozoic crust, shortened by the thrust faulting so beautifully revealed at the Red Rocks and the Valley of Fire, which were adjacent at that time, became distended. The pulling and stretching were so severe that the Red Rock and Valley of Fire areas became separated by about 100km. The Red Rocks and the Spring Mountains slid westward along the high and low-angle "normal " faults, and the "strike-slip" faults similar to the San Andreas Fault of California. In Southern Nevada, the great break in the earth's crust is called the Las Vegas Valley Shear Zone, a right lateral strike slip fault. The shear zone extends northward toward Reno, where it becomes the Walker Belt. Because the crust is being pulled apart from east to west, major north-south trending faults produced north-trending fault bounded mountain ranges and valley basins. The ranges and basins are also known as horsts and grabens.

Early exploration and mapping of the region by G.K. Gilbert led to the description that the mountains of the area resembled a herd of caterpillars crawling northward. The faults that bound the mountain ranges and valley basins cut very deeply into the crust, crushing and breaking solid rock. The faults provided many avenues for voluminous extrusive volcanic rock such as basalt and rhyolite. Economic minerals came up with the volcanic rock. Hot hydrothermal fluids also migrate up along the faults creating mineral deposits, hot springs and geothermal resources.

Cenozoic Life

With the disappearance of the dinosaurs, mammals were heir to the throne, and many species grew quite large before going extinct towards the end of the last major glaciation event, about 10,000 years ago. Thus, the Cenozoic is known as the "Age of Mammals." Perhaps the world's best record of these large extinct mammals is found in Southern California at the La Brea tar pits. The tar pits were formed when thick, gooey hydrocarbon rose to the surface along faults and fractures to collect in pools. After rains, the thick black goo became covered in water and appeared to be a good place to get a cool drink of water. The deadly scenario probably went like this: the vast herds of large herbivores came for a refreshing drink, a few ventured into the apparent lake and became stuck in the quagmire; packs of carnivores, seeing an opportunity, moved in for the kill, and they too became stuck. Scavengers, smelling death from miles around began circling overhead, but when they descended to dine and participate in the feeding frenzy, they were stuck. Then the oil worms moved in. Eventually the whole pile of decaying carcasses sank and their excellently preserved bones became part of the rock record as fossil hash. Many strands of food webs became entrapped and fossilized, providing a cross-section of the climax community 65 million years in the making, with such huge and vicious top carnivores as the saber tooth cat, the American lion, and giant dire wolf filling the niche vacated by *Tyrannosaurus Rex* of the previous dynasty.

The Pleistocene Epoch

When the Pliocene Epoch of the Cenozoic Era was ending around 3-2 mya, much of the world became locked by ice. We entered the Pleistocene Epoch, or the most recent Ice Age. For most of the last 2 million years, the Earth experienced successive formations of enormous continental ice sheets and mountain glaciers, followed by warm intervals. In order for snow to become a **glacier**, several things must happen. First, more snow has to fall than either melts or evaporates in a year, so that snow accumulates from year to year. Second, as the snow piles up, its weight crushes the lower levels. Sometimes it gets warm enough for some of the snow to melt, so it liquefies and gravity pulls it down. But then it refreezes, and so ice is made. Continue these processes for thousands of years until the ice is 100 meters thick. At that point, a threshold is crossed, and the ice has enough mass that gravity begins to pull it down. Once this happens, the ice is on the move, and a glacier is formed. Nothing seems to be able to resist the crushing, pulverizing force of a glacier, as whole mountains are crushed into scree and talus. Most of the highest mountains in Nevada and California had mountain glaciers in each ice advance of the Pleistocene. During each ice event, the cooler temperatures also allowed water to accumulate in the valleys between the mountains, many of which filled with lakes. Lake Lahontan in northwest Nevada was so large during most of the Pleistocene Epoch that it was the size of modern day Lake Erie. Pyramid Lake, Honey Lake, and Walker Lake are vestigial remnants of this once vast freshwater lake, up to 600 meters deep. Waters of Lake Bonneville lapped the shores in eastern Nevada. Throughout most of the Pleistocene Epoch, Bonneville was as large as modern Lake Michigan, but all that remains of Bonneville is its geologic record and the Great Salt Lake.

While during most of the Pleistocene the world was in an ice event, there were short periods (ca. 10,000 - 20,000 yr) of warming and glacial retreat. The giant mammals that developed in the Cenozoic survived these interglacial intervals, but the current interglacial interval is exceptional, and mounted a challenge that many of these species would be unable to overcome.

Around 11,000 years before present (BP), the glaciers around the world were receding again. The mountain glaciers in Nevada were melting and the valleys of the Great Basin began to fill with even more meltwater. Meltwater filled Death Valley until it became an enormous lake about 160 km long and 600 meters deep, that we call today Lake Manly. The ice cap expanding southward out of Canada and into the United States retreated, exposing the basins that are now the Great Lakes, carved by the glaciers from bedrock. The climate was changing and the ecosystems were responding. However, this time there was a species finally poised to take advantage of the interglacial interval.

The winds of change fanned the early flames of high technology. Arrowheads and spear points were skillfully chipped from stones carefully selected and coveted by early hunters and gatherers. Warming conditions continued, leading many of the dominant species to continue their spiraling decline. Many workers, especially the anthropologist Paul S. Martin, suggest that the human hunting technology and technique became so good that this time the large mammals succumbed to the interglacial. The succession of life forms proceeded, but this time without the astronomically loud sound of a huge deadly meteorite, nor the roar of a doomsday volcano. The global mass extinction of what we call the Pleistocene Megafauna (*mega-* = huge; *-fauna* = animals) was accompanied by the whoosh of a feather on an arrowshaft. Gone from North America are the great herds of camels, horses the size of dogs, huge woolly mammoths and mastodons, giant ground sloths, dire wolves, American lion, saber tooth tiger, and giant condor, to name a few. Left behind are the American bison, moose, elk, wolves, coyotes, grizzly bears, and California condors. The human inhabitants perfected simple structures such as tipis, and elsewhere they mastered more sophisticated stone condos called pueblos. The remains of Anasazi pueblos can be seen at the Lost City Museum in Overton, Nevada. Especially fine examples of old pueblos are found in Utah, Arizona, Colorado, and New Mexico. The Hopi and other pueblo-dwelling Native American tribes still live in these types of structures high atop mesas. Much has changed beneath the mesas.

CHAPTER 9

Nevada Energy Futures

J.D. Pellock

Introduction

From physics, a loose working definition of energy is "the ability to do work". The branch of physics known as thermodynamics, the study of energy transfers, has given us the Law of Conservation of Energy: Energy is neither created nor destroyed; however, its form can change. Essentially, when considering energy, our society as a whole performs a series of energy transformations in order to produce, maintain, and improve our lifestyles.

As an example, 100,000,000 years ago, a plant absorbed sunlight (energy) and carbon dioxide, and produced glucose (a rather dilute form of energy storage) and other necessary materials for the plant. After some time, the plant died, and it along with millions of pounds of dead plant material was covered over with dust, dirt, and rocks. There it lay for millions of years while pressure (the weight of the dirt on top of it) and heat from the earth turned the plant material into what we know today as crude oil. Crude oil can be burned directly (as in some oceangoing vessels), or it can be refined into heating oil, which is very similar to diesel fuel, and then burned. In either case, when a petroleum product is burned, the solar energy stored by plants as glucose, which was concentrated into crude oil, is then released as heat energy. When the heating oil is used to heat water in a boiler to produce steam, the pressurized steam now contains the energy stored in the crude oil. When the steam is used to turn a turbine, the energy stored in the steam is converted into rotational energy, or a form of kinetic energy. The kinetic energy of the rotating turbine shaft, if it is connected to an electric generator, will convert the kinetic energy from the turbine into electrical energy. The electrical energy, when delivered to our homes or workplaces, is then converted from electrical energy into a myriad of other forms of energy, depending on the needs of the person(s) or business and the device used. As a straight-forward practical example, if it is after sunset, and I want to read a book, I will turn on an electric lamp that will convert electrical energy into light energy (and also heat energy as well).

In order for any advanced technological society to operate and flourish, it needs to possess and use a constant and continuous supply of a concentrated form of energy of one variety or another. Either the energy source can be used directly, or through the use of technology and devices by conversion from one form of energy into another form. Unfortunately, as also stated by thermodynamics, every conversion is not 100% efficient, and there will be energy losses along the way. These losses include heat loss as in burning petroleum products in a boiler, friction which occurs with moving devices or parts of moving devices with moving parts, and electrical resistance which occurs in electrical wiring and systems. With proper engineering and design strategies, these losses can be made tolerable.

An unfortunate side effect of energy production, conversion, and use, as well as to a lesser extent its distribution, is its impact on the environment. Every state in the United States possesses unique opportunities and methods of making, producing, and storing various forms of energy. However, the final decision as to who, what, where, when, and why a particular form of energy will be used (converted) or exploited, is a decision which rests solely on the shoulders of the citizenry as individuals and in the forums of the town, city, county, state, and federal government.

The Energy History of Nevada

Nevada must import large quantities of fossil fuels (coal, petroleum products, and natural gas) for its existence and survival in the 21st century, as it has in the past. Electricity production in the state occurs through the use of hydroelectric generating facilities (Hoover Dam) and coal fired facilities such as in Laughlin, Clark Station outside of Las Vegas, and other sources as well. Most of the electricity from Hoover Dam is sold to the Los Angeles basin, so much of the electricity used in Nevada is either generated here or it is imported from out of state using the electrical grid from sources in the United States or even Canada. Energy use in Nevada, in the form of electricity and fossil fuel, has increased hand in hand with the increase in population over the last 30 years.

When population increases, so does electricity demand due to the simple fact that more people will need to have more electrical devices to live at a modern standard of living. The initial increase in demand for electricity nationally and in Nevada from the early part of this century through the late 1960s was phenomenal, due to the electrification over this period and the rapid design, development, and construction of energy using devices for consumers, business, and industry. Towards the end of the 1960s, solid-state devices started finding their way into the consumer market in the form of transistors. The use of solid-state devices dropped the power consumption of simple tabletop radios and stereos from hundreds or dozens of watts of electrical power to a few watts, or at most maybe a hundred watts (in most instances). This trend continued throughout the 1970s and 1980s with the increase and further development and use of new solid state devices. It even expanded to TVs, where the only tube is the CRT or picture tube. The typical tube type color 25 inch TV produced in the late 1960s or early 1970s consumed up to 1,200 watts of electrical power. The modern solid-state 32-inch TV in my living room consumes 125 watts of power. The TV became larger, has more features (such as stereo sound) a larger, clearer picture, and it consumes less power. For what it costs in energy to power a TV from the late 1960s or early 1970s, I can run approximately 10 new modern color TV sets.

While the change to solid-state technology went on during the 1970s and 1980s, there were also changes in the approach to energy use as well. Due to several energy "crises" and foreign policy difficulties in the 1970s and early 1980s, there was a strong national movement towards conservation or saving energy. The look at conservation was a national focus, with the aim not only at the design of consumer devices but also of homes and places of business. The outgrowth of this is the codified requirements of minimum R-values for home insulation, which will take less energy to heat and cool. It also includes the use of window placement and roof eave design for the use of sunlight for lighting and winter heating of rooms. These combinations of techniques slowed the almost exponential growth of electricity demand from the turn of the century through the late 1960s, to a slower steady increase in growth from the mid 1970s through the late 1980s.

Because of the slower growth in consumption due to conservation and devices with improved efficiencies, electric utilities did not see the need to build as many new generating facilities as they initially projected in the late 1960s or early 1970s. Nevada's power plants, as well as the nation's, had more than enough capacity to handle the growth of the 1970s and 1980s. Even expansion of existing plants was put on hold indefinitely.

The 1970s saw the start of a series of "energy crises." Broad-based environmental laws and regulations became more stringent, not only at the local level (city and county), but also at the state and national level. The additional burden for administration of the laws is costly, and in order to pay for the additional costs, a licensing fee was imposed on the licensee and the fees would be recovered ultimately from the end user. (This means the consumer will pay for the cost of licensure with increased electric rates.) For plants currently in operation, operating permits were required, often from more than one city, county, state and/or federal agency. The permits are costly, not only in terms of the outright fees for the permit, but also in terms of the paperwork required to obtain one, and the ongoing paperwork necessary to satisfy the permitting requirements and establish the groundwork for permit renewal.

As the 1970s gave way to the 1980s, building new facilities and expanding old ones was not the most viable option for reinvestment of profits, or acceptance of more corporate debt for an electric utility. Not only were generating facilities not improved, but electric grid infrastructure also was not drastically improved in this time span. Between growing environmental costs, a slow increase in consumer electric consumption, high construction costs, high interest rates, and increasing control of the electric industry from Public Utility Commissions and other branches of state and federal government, very few new plants or expansion of existing facilities occurred. Nuclear power during this period also became much more costly and one of the least politically expedient forms of energy to develop and use. Some forms of money for investment also dried up in the 1980s. The so-called "tax-shelter" investments of the late 1970s were no longer allowed under the Federal Income Tax Reform acts of the mid and late 1980s. "Tax-sheltered" money was the large impetus for building large commercial wind generators. Even as interest rates came down from 20% in the early 1980s, the costs associated with expansion or new facilities were too risky.

This policy of no large, significant increase in construction was followed through the 1990s. This led US electric utilities to purchase electric power from across our borders, namely Canada and Mexico. Because of a lack of electric grid infrastructure, electric power cannot always be shipped quickly and directly during periods of high electric demand. Unfortunately, this continuation is partially responsible for some of the large-scale blackouts in the western and southwestern states in the summer of 1996.

What Are the Costs of Expansion for an Existing Facility or Building a New Plant?

The 1970s, 1980s and 1990s saw the increase in regulations and, in some cases, control of electric utilities that made expansion or new facilities projects not the most viable option for a utility. In this section, I hope to illustrate what a CEO or president of an electric utility has to consider before committing to a project. It cannot be a quick decision, as there are many factors to weigh and consider before even digging the first shovel full of dirt.

Before a new facility can be built, or an existing one expanded, a planning study of the nature of the new facility must be performed. This will give a rough set of design criteria and operating parameters for it. After the initial planning study is performed, an EIR (Environmental Impact Report) must be developed by a consulting firm. An EIR will catalog and look at the plant and animal life impacted by the construction and operation of the facility, and can include such considerations as noise, heat, water use, air pollution, and water pollution. This study can take several months to several years to compile, and it can cost from $20,000 to a few million dollars for a particular site. If it is a new facility being constructed, several prospective sites may be looked at to search their suitability or unsuitability for a particular project. If a potential site is found to be unsuitable, it is crossed off the list for consideration. However, the cost for its study is non-refundable.

Once the number of potential sites has been narrowed down to approximately three, the EIR reports are made available for public inspection and comments. Once the general public and the pertinent regulatory and various local, state, and federal agencies have looked over the reports, public hearings are held for comments and questions. If serious questions or concerns are raised, or serious flaws are found, then more work is needed on the reports. This typically means more money will have to be spent, along with several months to a few more years of time is needed to compile the necessary data and amend the report(s). After the additions or improvements have been added to the original EIR reports, the public is allowed to inspect the reports and attend public review meetings. Sometimes this process can go back and forth several times; other times, it can sail through easily on the first try. It varies by project type, size, and location.

Once the EIR reports are approved by the appropriate governing agencies and the public, the "go-ahead" for the project is granted to the utility. At this point, the land, if not already purchased by the utility, is finally bought, and final construction plans and details are finished. Local and state agencies will look over the plans and specifications to ensure construction building codes are met and exceeded. Construction permits from the local jurisdiction(s) must be issued, and also add to the cost. Once construction plans are formalized, construction can begin.

If during construction an endangered species of plant or animal is found, or a skeleton or ancient human artifacts are unearthed (signifying an unknown burial ground), then the whole project must be stopped immediately. In the case of an endangered species, typically further EIR report work must be performed, along with public review and meetings. This will take time and money. A determination then must be made concerning what to do about the endangered species, with typical options being the species might be relocated if it can be, or the project is terminated. With a burial ground, the options can vary. Sometimes the body or bodies can be relocated and buried elsewhere, other times, the site must be dug up and catalogued by an archeologist or a team of archeologists. Often construction can resume after a period of time, or sometimes not at all.

Depending on what phase of construction the building was stopped, startup may be inexpensive, or it may have become very expensive. For example, agreements with suppliers and subcontractors may have to be renegotiated, or fees paid for storage or non-purchasing or payment for goods or services. Cement for foundations, footings, and floors are best poured continuously. If a concrete pour is stopped and then started again, the resulting cold joints will give a significantly weaker cement product. Workers may have to be paid wages while just standing around, not working until a course of action is figured out. If the site is not occupied regularly during the shutdown period, the whole site is prone to vandalism. If the project can be restarted, the cost for the project will increase. If partial concrete pours were made instead of continuous pours, what was poured will either have to be removed, or parts of the project re-engineered and/or redesigned to work around the partial pours. This is an expensive process in terms of capital and time.

In several cases, construction was completed on the facility, however a class action lawsuit against the facility caused it to be completely shut down, mothballed, and then abandoned. In some instances, the project was a total waste of time, effort, and capital for the electric utility. A summation of the possible expense burden is in Table 9.1.

As a CEO or president of an electric utility, a very simple project can balloon into a very significant expenditure, without any guarantee of a return if the project has to be scrapped.

Table 9.1. Cost range of a project for expansion of an existing facility or a new facility depending on the size of the facility and the level of difficulties encountered on the road to its completion. Please note that the low dollar figure is for a small project without any significant obstacles or impediments, while the largest dollar figure is for a large project that has met very significant problems along its path to completion.

Expense	Cost
Initial EIR reports	\$20,000-\$5,000,000
Facility design and engineering costs	\$100,000-\$5,000,000
Initial Construction Costs	\$500,000-\$100,000,000
Follow-up EIRs	\$20,000- \$3,000,000
Construction Stop and Starts	\$1,000,000-\$15,000,000+
Facility Re-Engineering	\$10,000-\$5,000,000
Legal Fees	\$20,000-\$2,000,000+
Licensing and Permitting Fees	\$12,000-\$2,000,000
GRAND TOTAL FOR THE PROJECT	**\$1,652,000-\$127,000,000+**

Fuel Issues

All forms of commercial electricity generation need some sort of fuel, whether it is a fossil fuel, a waste by-product (shredded tires or agricultural wastes) for a co-generation facility, or water stored in a lake behind a dam. All facilities for commercial electricity production can have the same construction headaches as outlined above, although some fuel sources have fewer hurdles to overcome than others. Each fuel or fuel combination has its own unique set of advantages and disadvantages, and has to be considered on a case-by-case basis.

Coal

Coal is by far the most abundant fossil fuel in the continental United States. It is relatively cheap and abundant. Its cost is fairly stable, and only the cost of shipping varies. It also contains a great deal of energy per pound of material. Coal is easily shipped and stored in the open. Mining is principally performed through the use of explosives and strip mining techniques. This has drastically cut back the incidence of mining accidents and the occurrence of "black lung".

The drawbacks to it are numerous, however. As was stated above, most coal is strip mined, which can be very hard on an ecosystem. Current federal regulation forces mining companies to place some profits into a trust fund, which will ensure that the land is reclaimed after mining operations have ceased. There are some parts of Kentucky where mining, in its classic sense, begins with tunnels and automated equipment. In all of its forms, coal contains sulfur and sulfur containing compounds. When it is burned, the sulfur and sulfur compounds produce sulfur oxides, which will combine with water in the atmosphere and will produce sulfuric acid, a major component of acid rain. The lowest concentrations of sulfur are found in hard coal, and for all practical purposes soft coal, has been banned from use. Coal will also produce large quantities of soot and ash. Whenever coal is burned, scrubbers and dust precipitators must be used to remove air contaminants. The higher the sulfur content of the coal, the more expensive scrubbing becomes. Because of the issues of sulfur, airborne particulates, and nitrogen oxide emissions, coal burning plants are moderately difficult to build or modify.

Oil

Oil, while it is a very condensed form of energy, has a series of unique problems, mostly associated with cost or the cost/pricing structure of it. Since the US imports so much foreign oil, its price also is affected by many external factors besides the actual cost of drilling an oil well and pumping the oil out of the ground.

Firstly, foreign crude oil production quantity maximums and minimum pricing is set by OPEC (Oil Producing Exporting Countries). This is a bare minimum that oil will sell for on the open market. Also, crude oil is sold as a commodity, and as such its price can vary greatly due to purely psychological factors beyond supply and demand issues and OPEC (a perceived demand or a perceived shortage of oil can cause the market to increase asking prices artificially). Oil prices will also vary seasonally in the commodities market due to perceived and actual supply and demand issues. For example, if the winter in the northeast is harsh, more heating oil is needed, and so demand increases for more crude oil. This causes the price to increase due to the limited supply being doled out by OPEC countries. Secondary to market factors, events in the Middle East can also have a marked effect on oil prices. When there is instability in the Persian Gulf, oil prices will increase. With all these issues surrounding oil pricing, it is very difficult for an electric utility to lock in a long-term price or pricing structure for the facility. Diesel/heating oil is also the raw starting material for unleaded gasoline manufacture. Increased demand for gasoline will drive up the cost of heating oil, which is also used in oil fired electrical power plants. This will also increase the cost of producing electricity. These factors typically cause the cost of operating the facility to fluctuate a great deal, oftentimes making oil produced electricity quite costly, or at the very least much less profitable.

Oil, like coal, can have its emission problems. Crude oil often contains significant quantities of sulfur, and it must be removed during the refining process. This adds cost to the fuel. Burning oil will also produce greenhouse gasses, and sometimes soot as well. The soot and sulfur emissions can be dealt with using off-the-shelf technology by using electrostatic precipitators and scrubbers.

Natural Gas

Natural gas is typically the preferred fossil fuel for electricity generation. While its cost can fluctuate seasonally due to seasonal demands of the weather for heating, it is in fairly abundant supply in the continental US. With an extensive pipeline system, it is easily transported safely. If its combustion process is monitored carefully, there will be only minimal nitrogen oxides and sulfur oxides produced from its use. No particulates are produced. Typically, no scrubbing or particulate precipitators are necessary.

Co-Generating Facilities

Co-generation facilities use two or more fuel materials to produce heat, which in turn produces steam in a boiler. The combination is a fossil fuel burned with a waste material such as an agricultural waste (almond hulls) or scrap (old tires). One of the most common combinations is natural gas and shredded tires. Location can be critical for the economical shipping of a waste material, especially if agricultural wastes are used. Burning this kind of a mixture causes unique measures to monitor the combustion process and to ensure emissions are not exceeding the permitted levels. Often scrubbers and electrostatic precipitators are used to trap airborne pollutants. Greenhouse gasses are emitted, as well as a particulate contaminants and possibly odors.

The transportation and storage of the waste materials can cause problems as well. For example, in Texas, tires were imported from Japan for a co-generating facility. Inside the tires was water which contained tiger mosquito larvae. The larvae matured, and now have started to breed and multiply in the US. For plants designed to use agricultural waste, the hulls from agricultural crops have to be trucked into the facility, and they can sometimes absorb water and ferment or start to rot (foul odor). Wet materials also hinder the combustion process.

Water for Hydroelectric Power

The only fuel requirement for hydroelectric power is a body of water that can drop some distance to turn a turbine. The building of the dam and the subsequent formation of a lake behind the dam causes many problems. One of the first requirements for a large lake is the purchasing of a large quantity of real estate. Since real estate is in constant demand, and they are not making any more of it, there can be quite a large expense in trying to acquire it. If the desired land is inhabited and has homes on it, besides purchasing the land, the people have to be relocated. Depending on the methods used, this is an expensive and painstaking process at best. Often, since a branch of the government is putting in the dam, **eminent domain** can be used. This is a legal process in which land is declared to be underutilized and not making a reasonable profit. This enables the state or more likely the local government to force the sale of the land. The declaration of under-utilization causes grief, and also the price set or determined for the value of the land often causes disputes between the owner and the government. However, it is often seen as an abuse, and the branch of government trying to put in the dam may find itself in courts for several years. This results in paying much more than the fair market value for the land, along with rather large legal fees for both parties of the suit.

The EIR for a large hydroelectric project will typically be expensive and it will usually describe the destruction of several natural animal habitats and ecosystems. Large dams also will have a finite life in terms of silt retention, due to the rivers and streams emptying into the lake, carrying not only water but also silt. Slowly over time, the lake fills with silt, either making the dam useless, or requiring the lake bottom to be dredged. Lake Mead, behind Hoover Dam, will only have a predicted lifespan of 350 years. On the positive side, hydroelectric power does not produce any greenhouse gasses or emissions. In some cases, new and/or different ecosystems may be produced with a large hydroelectric dam project; however, the public will have to weigh the risks versus its benefits.

Large dams and lakes are difficult to nearly impossible to get approval for. However, this does not totally preclude new hydroelectric projects. Smaller mini-dams typically have an easier time winning public approval. Several small dam projects can be just as effective, or more effective, than a single large project. For example, in California, several irrigation districts have built small hydroelectric projects on their water control **weirs**. In essence, they already have a dam built, and need a minimal EIR for the building of a generator house. For a water drop of as little as fifteen feet, a small turbine can be used to generate 10-40 kW/h (kilowatt hours) of electricity. For several small dams or weirs, this can add up to quite a lot of electricity generated from an irrigation district's distribution system. The power can be used by the district itself to operate its pumps, and/or excess can be sold to the local electrical utility. Two large irrigation districts (Modesto Irrigation District [M.I.D.] and Turlock Irrigation District [T.I.D.] operate their own electrical utility for their respective cities, and their rates are significantly less than most neighboring electrical utility companies. They also have older, medium-sized dams.

Nuclear Power Plants

The nuclear industry is dying a slow death by attrition. No new plants are slated for construction due to the very high cost of design, construction, operating expenses, extensive EIR requirements, and the public image problems associated with the nuclear power industry. Nuclear reactors also need large quantities of water for cooling and operation. Despite the criticisms against it, nuclear energy produces no greenhouse gasses, zero air pollution emissions, and no radiation emissions. The design and use of the water-moderated reactors in the US are pretty much a fail-safe design, and the reactors will not blow up producing a "mushroom cloud." In fact, it would not explode. However, it could melt the internal parts together.

Nuclear waste is a constant issue for southern Nevada because of the Yucca Mountain Nuclear Waste Repository. Most waste from nuclear reactors can be recycled and reused. However, because of a Presidential Executive Order signed by then President Jimmy Carter, spent fuel must be buried, not reprocessed (recycled). This turned a 70%-80% recyclable resource into garbage. With additional research using newer laser separation methods, better than 90% of spent fuel rod material is recyclable.

Alternative Forms of Electricity Production

Due to unique features of Nevada, it is often suggested that so-called "alternative forms of electricity production" should be used. This suggestion is based on high expectations of the technology's high performance and low cost. Alternative methods work; however, trade-offs associated with their use often are not mentioned or are glossed over. Some of the methods (especially solar) may also be useful for helping utilities get over peak hour usage humps.

Solar (Photovoltaic Cells)

It stands to reason with all the sunny days in southern Nevada that using photovoltaic cells (solar cells) would be a "no-brainer." Solar cells could be used for large scale, commercial applications, or to prevent shortages during periods of heavy use. The simple reason is that solar electric generating systems are too costly for the ten hours a day maximum of electrical output. (What do you do at night?) High efficiency cells are expensive to purchase and they need a significant amount of unobstructed space to generate electricity. For a commercial application, this means a large amount of real estate is needed for the facility. While they can be constructed for a "set and forget" type of low-maintenance installation, they still need room. If industrial land cannot be obtained, it is suggested that systems be installed in the desert. However, the federal government claims title to most of the desert, and purchasing that land outright requires, literally, an Act of Congress.

Wind

Wind generators work great as long as the wind is blowing. If it is not blowing, the wind generator is useless. If the wind generators are placed in strategic locations (i.e., places with lots of wind, such as mountain passes), they can work very well. Other drawbacks have to do with maintenance and noise problems. Since they are mechanical, they need to be checked periodically, which can be quite expensive. Regular readings on how much power they have generated must also be done regularly (to bill the utility). The driving of vehicles can disrupt wildlife in sensitive areas. Also, the mechanical noise of the blades and the generator can be annoying to wildlife. Some designs are not very "bird-friendly," and from anecdotal information, maintenance personnel have referred to birds as being "cuisinarted" by the blades.

Geothermal Energy

The use of geysers and hot vents for steam production has been around for many years. If a natural vent exists, it would be fairly inexpensive to start out, at least from the surface. But looks can be deceiving. Naturally occurring steam vents tend to produce steam with a fairly low steam temperature and a very high water hardness and brine concentration. High brine concentration, coupled with elevated temperatures, causes a great many corrosion problems. It causes problems not only for the plumbing between the vent and the turbine, but also the delicate turbine corrodes as well. The corroding of the turbine blades means the blades have to be removed from service and repaired on a regular basis of approximately every eighteen months. The process is expensive and has to be performed in a separate facility, often hundreds of miles away. The low temperature of the steam limits the efficiency of the turbine, and its efficiency will not be anywhere as near as high as that of a turbine operating off a boiler fired with fossil fuels. This translates into producing less energy or electricity for a given size of turbine.

Homestead Energy or D.I.Y. (Do It Yourself)

Developing your own electrical system or purchasing an installation for your house has distinct advantages. You can be off of commercial power mains and never again be bothered by a black-out or a brown-out. However, you need to be aware of what can and cannot be done with your system. In southern Nevada, home air conditioning is a must in the summer. If you have a swamp cooler, most photovoltaic systems capable of 3 kW/h of electrical production could handle the demands of a typical household, provided you have gas appliances. If you have a refrigeration type of air conditioner, you will be sorely out of luck, since they require 8-20 kW/h or more of electricity. In the next year or so, Southwestern Gas will recommend home heating and cooling systems operated with natural gas, which may make home generated power much more attractive. If you go this route, you will have to become very familiar with your system and at least at this point in time, you may have to piece a system together from several different component manufacturers. Not only will you also have to maintain the system yourself, you also may have to use more than one type of generator system and/or have a fossil fuel back-up.

Obviously, home generated power systems are not for everybody for a number of reasons. As with any added device, there will be maintenance and maybe construction activities associated with them. Not everyone has the ability to design, build, buy, assemble and maintain a system. The cost for a photovoltaic system could range between $10,000-$20,000 for a typical house installation. With most new home construction in southern Nevada, you will have a homeowner association and covenants and restrictions that limit what changes can be made on a property. Some even specify what can be placed in a backyard! Additionally, not all roof elevations lend themselves to the installation of solar arrays on them, and tile roofs should not be walked on or penetrated for wiring or mounting.

Practicality of Alternative Energy Sources

In order for most alternative energy sources to work, tax incentives of some sort have been used to subsidize the effort. This means that the cost is spread out over the whole taxpayer base; in essence, the builder/owner is not paying for the system but all taxpayers are. What is really needed for alternative energy sources to succeed (especially solar energy) is a massive technological breakthrough to improve cell efficiency. Possibly a breakthrough in room temperature superconductors would assist in making electrical wiring and devices more energy efficient, thus requiring much less power to operate. Improvement in packaged turn-key systems for solar and wind would also help as well.

The Hydrogen Economy

A fairly new proposal is the switch from a fossil fuel economy to a hydrogen based economy. The logic I cannot disagree with greatly. The proponent's first point is oil is very expensive not only in terms of purchasing cost upfront, but also in terms of foreign policy and environmental costs. We have had a major war in the Middle East, and the possibility of a second war after 9/11 is likely. Saddam Hussein and Osama Bin Laden have derived much of their funding from foreign oil purchases, which they would not have as much access to if we were on a hydrogen-based economy. Proponents include several large oil companies, the US Department of Energy, and several automobile manufacturers.

The proponents suggest that solar and wind machines produce electricity which can then be used to split water molecules into hydrogen and oxygen molecules. The hydrogen can then be collected, compressed, and shipped via pipeline like natural gas. It has even been suggested that hydrogen could be liquefied and the pipeline used not only to transport cryogenic hydrogen, but also to place a copper conductor inside the pipeline to act as a high voltage, super conduction power line. This scheme still needs a great deal of planning, coordination, and engineering work before it becomes practical. However, programs are underway to have significant movement towards a hydrogen economy by 2015 or sooner.

Hydrogen, as a fuel, has several obstacles to overcome before it can have widespread use. Even though hydrogen production and transportation has an extremely high safety record for a long period of time, the public is leery of using it because of the images of the Hindenburg zeppelin burning. The fact is that most of the deaths came from people jumping out of the dirigible, not as a result of the hydrogen burning. The actual hydrogen flame burned upward and quickly. As a motor fuel, it would be just as safe or safer than gasoline in an accident, since it will move upward and diffuse quickly in the atmosphere.

Hydrogen gas, because of its small size, tends to diffuse into the container material holding it. This will make many metals brittle over time (hydrogen embrittlement). The current solution is to carefully choose metals that will be in contact with hydrogen for storage and transportation, along with a container testing and inspection program. Testing and inspections must be performed regularly. Pumps for compressing the gas and moving it through a pipeline also have to be made with materials that will withstand hydrogen embrittlement. Automotive engines and engine parts will also have to be developed to withstand higher combustion temperatures and tolerate hydrogen embrittlement. The obvious materials for storage tank construction are either fiberglass or plastics, however a plastic that can contain hydrogen without failing does not exist yet.

Liquid hydrogen is a cryogenic fluid, and it exhibits very interesting properties. It is a very good insulator, and a conductor at room temperature will become a superconductor in liquid hydrogen. Due to its low temperature, it must be stored in a container with extensive insulating schemes. Otherwise it will boil off and vaporize quickly, turning into hydrogen gas. Most tanks that hold liquid hydrogen must be made of carefully selected aluminum or stainless steel alloys. For NASA's space plane program, plastic liquid hydrogen tanks are being developed. The latest results are for a plastic that appeared to fail due to a small manufacturing defect.

The hydrogen economy proposal has many very appealing points, and it could use a source of electrical generation that is not capable of running 24/7. The downsides to the proposal are the needs for many basic engineering solutions to be worked out, standards to be developed, and testing methods and protocols to be determined for tanks, automotive plumbing, pipelines, and dispensing. Currently no considerations have been given to brine disposal from the production of hydrogen from water. From the pollution standpoint, there would not be the greenhouse gas emissions as there are with any fossil fuel, and the horsepower should not suffer from automotive engines. However, nitrogen oxides and ozone will still be produced by automotive engines, which is not mentioned in most of the literature. It may also help to reduce or eliminate some foreign policy problems for the US as well.

Where Do We Go from Here?

Conservation techniques and improved designs of devices and homes has helped to extend our energy sources. However, with an increasing population base and an increasing need for energy, fossil fuels will have to be continuously imported into Nevada. Coal is needed for existing power plants, and the new power plants on the drawing board will run on natural gas. Co-generation now appears unlikely. Nuclear energy in the near foreseeable future will not be developed in Nevada for consumer power. Fusion will not become viable until superconductivity at room temperature is achieved to make very large magnetic fields for use in a Tokamak reactor. Fusion will produce radioactivity and waste, but not as much waste as from nuclear fission.

Oil is a rather interesting beast. Besides OPEC and the other factors mentioned previously, a major factor effecting supply is the fact that no new oil refineries or oil processing facilities have been built in this country since 1976. Those refineries still operating do so at typically +96% of rated design capacity 24/7, for most of the year. The only time maintenance can be performed is in September, after school starts, due to lower gasoline demand. Crude oil refining is becoming more complicated for diesel, heating oil (which is about the same as diesel fuel), and gasoline, due to regulated lower sulfur contents in fuels. Gasoline in particular will increase in price since the current regulatory framework requires several different oxygenation formulas for various regions of the country. Some of the formulations are needed in very small markets, which drives up the cost of gasoline markedly for that region. As was stated previously, diesel/heating oil is also the raw starting material for unleaded gasoline manufacture. Increased demand for gasoline will drive up the cost of diesel for transportation and shipping, while increased prices for heating oil will increase the cost of producing electricity in oil fired facilities.

The hydrogen economy could work quite well for a state like Nevada. Hydrogen can be produced by a myriad of methods using intermittent alternative noncommercial forms of electrical generation. Currently, solar and wind power look very attractive. However, a great deal of engineering needs to be done before it becomes practical, and water is a very precious commodity in most parts of the state.

What can we do, as individuals? As with most topics, the energy future of Nevada is up to us, the citizenry. Each one of us can conserve energy by using simple conservation techniques, from simply turning off a light to adopting the use of energy saving devices—fluorescent lights instead of halogen or incandescent lights. We can be come educated on the issues by reading books, "Scientific American" for the latest technological updates, discussing the issues with others, and watching documentaries of PBS, The Discovery Channel, The learning Channel, and the History Channel to find out what is happening, what could happen and what has happened in the past. In fact, sometimes there are interesting forums or subcommittee meetings dealing with energy on CSPAN, and you can hear our various national political leaders complete speeches, discussions, interviews, or news conferences — not a "talking heads" attempt to decipher the news for us (typically incorrect or quoted out of context). In short we all have to make a conscious decision to get involved, dig deeper into an issue, and take control, or else let someone else make a decision for you. Typically, that will be a politician who may not be concerned with a long-term vision. The future offers several very interesting possibilities. It is up to us to develop and apply them, or let them go by the wayside.

CHAPTER **10**

Nevada Solid Waste and Recycling

Denise Signorelli

Nevada Solid Waste

Nevada has enormous expanses of unpopulated desert, 60 million acres of which is managed by the federal government (see Chapter 4). There are only 18.2 Nevadans per square mile, compared to 79.6 as the National Norm and 401.9 New Yorkers per square mile according to the 2000 Census (see Chapter 2). Thus, Nevadans have less trash and plenty of space to put it. In addition, the desert is often called a "wasteland" and the common perception is that the land is good for nothing else besides the storage of waste. Therefore, Nevadans think they are exempt from the concern of many easterners; *where do we put all the trash*? This has lead to complacency, but the federal government has enacted a few statutes so that Nevada and other states are required to reduce waste and improve the safety of waste sites.

The volume of solid waste in Nevada is considerable. The federal government has set a standard for every state to recycle 25%. In Nevada during 1999 (the most recent available statistics), the state had a recycling rate of 11% and Clark County a rate of 8%. That means that roughly 90% of the waste generated in our state goes directly to the landfill.

Landfills in Nevada have been the primary focus of the Nevada Department of Environmental Protection (NDEP). Over the last several years, many landfill sites have been "upgraded." Sites that were not legal, like open dumps, have been closed, and legal sites have been improved with liners, fencing, landscaping, and other improvements.

The issue of available space vs. usable or reasonable space in Nevada is not currently a major focus of the state government or NDEP. A conference on waste and recycling for the state held in Las Vegas among federal, state and local government officials and interested parties was cancelled in May 2000 for "lack of interest." As technology improves, and humans move further into the desert *wasteland* it is likely that landfill space will become more valuable and the monetary value of desert land should encourage behavior that is more responsible.

In the two high population centers of Nevada — Clark and Washoe counties — landfills have already become neighbors that no one wants. The idea of NIMBY, or "not in my backyard," has caused large landfills to be relocated to areas where the population is less likely to fight them. In Washoe County, several landfills have been closed because they are unwanted neighbors. Politically savvy Tahoe residents had the dumps closed, citing the bad smell and the potential for leaking dangerous waste.

Hazardous waste is under close Federal and State scrutiny. Any municipality with 100,000 people or more is required by Nevada statue to provide curbside recycling and household hazardous waste disposal. Municipalities with 25,000 to 100,000 residents must provide drop-off centers for recycling and hazardous waste. Any place that has less than 25,000 residents is exempt from these statutes. In compliance, Clark and Washoe counties both have drop-off sites for household hazardous wastes that are provided as part of the solid waste disposal contracts. Carson City, though not required, also provides a hazardous waste disposal dump. What this means is that the smallest communities are the most likely to have substandard landfills, and have no alternative method for disposing of household hazards like used motor oil, antifreeze, paint, and batteries. Consider that many rural areas depend on groundwater for their drinking water and you can only hope that their waste level is so insignificant it warrants no further remediation. What do you think? Is this sound, economically feasible policy?

In Nevada, the hazardous waste that receives the most publicity is spent nuclear fuel. Though Nevada does not have a nuclear power plant, it has the only site being studied for permanent disposal of at least 77,000 tons of spent nuclear fuel. Is this sound policy or an accident waiting to happen?

In 2001, Nevada became the "most polluted state in the nation," a label that state politicians were loathe to admit. This label was given due to the high volume — 3 billion pounds of toxins released into the air, water, and soils — primarily from hard-rock mining. These mine tailings are generated each year in our state. Since these tailings are considered wastes with the potential of polluting soils, air, and water, they gave Nevada the number one spot on the polluted list. Politicians were unhappy, because mine tailings were added to the definition of pollution in 1997. President Clinton expanded the pollution list in 1997 to include mine tailings and six other kinds of pollution under the "citizens' right to know." Yet, some of these tailings contain cyanide used to extract gold or high levels of heavy metals and arsenic, and when the chemicals leak into the water table, the water is not safe to drink. Some argue that mine tailings are natural and thus should not be included under the definition of pollution.

What do you think? Are soils and rocks dug from under the surface and spread on the surface pollution?

The best low cost alternatives to disposal of solid waste, whether they are hazardous or not, is recycling. Many household wastes such as glass, plastic, paper, used motor oil, and batteries can be recycled. Even spent nuclear fuel can be recycled (see Chapter 9), greatly reducing the volume needed in landfills.

Illegal dumping continues to be a problem throughout the state. Because the desert biome is fragile and decomposition is very slow, the results of dumping centuries ago are still visible. Local and state authorities offer rewards to anyone who reports an illegal dumper.

Compared to national standards, the 9.4 pounds of trash Nevadans throw away each day is nearly three times the national average. What makes Nevadans so trashy? The prevalent logic is that much of our trash is generated by tourists, and they have little interest in cleaning up our state because they are on vacation and don't live here. In addition, if you look at states that have high recycling rates, they all have a "pay as you throw" program. This is a method of charging for waste disposal dependent on the amount thrown away.

Recycling

Imagine that recycling is a convenient curbside activity for you at a nominal cost. Then to discard of other items, only "approved" bags or bins could be used. The fees are sliding; each bag costs $5 each and only the approved bags are handled. Each 10-gallon bin costs $50 a month to empty, and anything that does not fit in a bag or the bin will not be taken. This is an economic incentive to reduce the trash stream.

Consider also the amount of lawn waste in our trash stream. No county in Nevada has a working composting facility. Thus, tons of grass clippings, tree trimmings, and weeds from homes, apartments, businesses, and parks are buried in landfills. Many municipalities now have convenient pick up or drop off for these kinds of naturally recyclable wastes.

The United States generates nearly 90% of the Earth's human waste, discarding 62,710,000,000 tons of waste in 1998. By comparison, China generates about 300,000,000 tons per year with a population of 1.2 billion. Remember that as a national average, the US recycles 25%; in Europe, Austria and Germany have a recycling rate greater than 50%, and four nations are at exactly 50%. In Nevada, the recycling rate is 11%!

Clearly, Nevada needs ways to improve. There are two general methods to decrease waste: one is by pushing and the other is pulling. To push people to reduce their waste, two methods are commonly used; (1) encouraging recycling and reuse by making it easy and cheap if not profitable, and (2) increasing our ability to get energy from waste (burning wastes or using methane produced during decomposition as fuel). Pull methods are strong ways of encouraging waste reduction. Pull methods include strong policies and legislation mandating compliance and economic disincentives to waste production.

In Nevada, state statute provides for a 24-hour toll free number to support waste reduction and recycling: 1-800-597-5865.

You can think globally and act locally by your actions. Consider this list of suggestions to reduce your personal waste stream.

1. REDUCE. Don't buy it if you don't intend to keep it.
2. REDUCE. Consider excessive packaging a reason to choose a different brand or product. For instance, leave the individually wrapped cheese slices and choose slices that are packaged together (90% reduction in packaging).
3. REDUCE. Encourage your workplace or school to go "paperless" or at the very least commit to buying recycled products.
4. REDUCE. Complain to businesses you frequent if the products they give you are bad for the environment or stop buying their products. For instance, polystyrene releases CFCs when it degrades and takes a very long time to decompose. Even McDonald's, not known as an environmental watchdog, stopped packaging their sandwiches in Styrofoam and uses paperboard or paper instead.
5. REUSE. Reuse everything. If you wash and reuse plastic sandwich type bags just once, you've reduced your demand for that petroleum-based product by 50%.
6. REUSE. Buy reusable products instead of disposable.
7. REUSE. See if your community has a reuse center for discarded building materials, office supplies, and furniture. Donate to it and take things from it as needed.
8. RECYCLE. If available, recycle all you can at your curbside. Make it a goal to see how little you can send to the municipal landfill.
9. RECYCLE. Give things to charities that you no longer use instead of throwing them away. Have a garage or yard sale; your junk is another man's treasure.
10. RECYCLE. Close the loop by buying products that are recycled or have recycled content.
11. RECYCLE. Save things like motor oil, and batteries, and take them to an establishment for recycling.
12. RECYCLE. Start your own composter to recycle yard and kitchen waste. If this isn't practical, lobby your government to do it locally.

Nevada recycling is primarily the responsibility of the Nevada Department of Environmental Protection (NDEP). However, five nonprofit organizations also are involved. A great deal of the current focus is to create local markets for recycled products. The government's efforts are funded by a $1 per tire disposal fee.

In the future, it is likely that the state's policy will be stricter with businesses. Now the policies are aimed at the residential sector. The commercial sector must bear some responsibility to recover recyclable materials if Nevada hopes to meet the federal goal of 25%. In addition, government agencies may soon be required to "buy recycled." Residents will continue to be encouraged and educated about the opportunities to reduce their waste.

CHAPTER **11**

Global Climate Change Research in the Nevada Desert

Lynn K. Fenstermaker, Therese N. Charlet, Travis E. Huxman, James S. Coleman, Robert S. Nowak, and Stanley D. Smith

of the Nevada Desert Research Center*

*A Global Change Experiment in the Mojave Desert

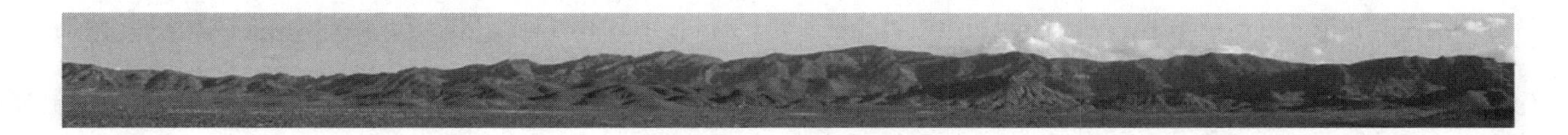

The mission of our center is to gain a better understanding of the ecological function and structure of the Mojave Desert at all scales, and how desert ecosystems might be impacted by global climate change. These types of studies are important because approximately 40 percent of the earth's terrestrial surface is arid or semiarid, and more land is undergoing **desertification** each year. The Nevada Test Site is an ideal location for ecological studies because of the accessibility of large tracts of pristine land that have been protected for at least 50 years, and it encompasses several plant communities along an elevation gradient, as well as Great Basin-Mojave Desert transitional communities.

*A collaborative research effort of the University of Nevada Las Vegas, University of Nevada Reno, Desert Research Institute, Brookhaven National Laboratory, and the Department of Energy

Currently two large projects are underway at the Nevada Desert Research Center. These projects are the Nevada Desert FACE Facility, a study that examines the effects of elevated CO_2 on the Mojave Desert ecosystem, and the Mojave Global Change Facility, to examine the effects of other predicted global change factors, namely nitrogen deposition, crust disturbance, and increased precipitation. In addition to gaining an understanding of the possible responses of the Mojave Desert to global climate change, the current research will also provide a better understanding of the Mojave Desert biosphere at molecular, physiological, and ecosystem levels, as well as soil dynamics and the interrelationships among plants, herbivores, and insects.

This facility is located on the Nevada Test Site near the northern ecotone of the Mojave Desert. Current landholdings assigned to our facility include two square kilometers between 36° 46' 30" North, 115° 57' 45" West and 36° 45' 20" North, 115° 59' 15" West. This area is a fairly homogeneous area on a broad gently sloping bajada (coalesced alluvial fans). The vegetation is characterized as desert scrub, with small-statured shrubs that provide less than 20% total ground cover.

Nevada Desert FACE Facility

The Nevada Desert FACE Facility (NDFF) is a state-of-the-art research facility designed to study responses of an undisturbed Mojave Desert ecosystem to increasing atmospheric carbon dioxide (CO_2). Desert FACE is a cooperative project led by the University of Nevada Las Vegas, University of Nevada Reno, Desert Research Institute, and Brookhaven National Laboratory, with major support from the US Department of Energy and the Nevada Test Site.

The Nevada Desert FACE consists of nine study plots, each 23 m in diameter; three FACE rings at elevated CO_2 concentration (550 ppm), three FACE rings at ambient CO_2 concentration (*i.e.* control rings at ambient CO_2 concentration of 360 ppm), and three non-blower control plots. The array of study plots is located on a broad alluvial fan in vegetation that is dominated by *Ambrosia dumosa* (white bursage), a small drought-deciduous shrub, and *Larrea tridentata* (creosotebush), an evergreen shrub that reaches over 1 m in height. Other important shrubs include the drought-deciduous *Lycium andersonii* (Anderson's wolfberry), *Lycium pallidum* (pale wolfberry) and *Krameria erecta* (desert ratany) and the evergreen *Ephedra nevadensis* (Mormon tea). Abundant perennial grasses are *Achnatherum hymenoides* (Indian ricegrass), *Pleuraphis rigida* (big galleta), and *Erioneuron pulchellum* (fluffgrass). Up to 75 annual species may occur depending on rainfall, including the exotic annual grass *Bromus madritensis* ssp. *rubens* (red brome). Soils at the site have a well-developed microbiotic crust made up of bacteria, algae, mosses, and lichens that makes atmospheric nitrogen available to organisms by a process known as fixation.

What Is FACE?

FACE stands for Free Air CO_2 Enrichment. FACE technology allows us to increase the CO_2 content of the air around natural vegetation while keeping all other environmental and ecosystem conditions natural. FACE technology avoids the many problems that are associated with elevated CO_2 chambers and greenhouses, and allows direct extrapolation of ecosystem responses to future global environmental conditions. Thus, Desert FACE provides the capability *today* to study environmental conditions of *tomorrow*. A network of FACE experiments around the globe is conducting similar studies in other ecosystems. Links to other FACE facilities can be found at our website, http://www.unlv.edu/Climate_Change_Research

What Is the Concern about CO_2 Levels in the Atmosphere?

Current atmospheric CO_2 levels are 25% higher than preindustrial levels. With the present rate of increase (Figure 11.1), atmospheric CO_2 will be approximately 550 ppm in the year 2050 and will double current levels by about the year 2070. Many plant species have been shown to be extremely sensitive to increased atmospheric CO_2, but the cumulative effects on plant processes and on ecosystems are not clearly understood. Elevated CO_2 is likely to significantly impact the species composition, **biodiversity**, water availability, vegetation cover, **surface albedo**, and a host of other characteristics of desert ecosystems that may result in a substantial change in the **physiognomy**, functioning, and land-use patterns of this prevalent ecosystem type.

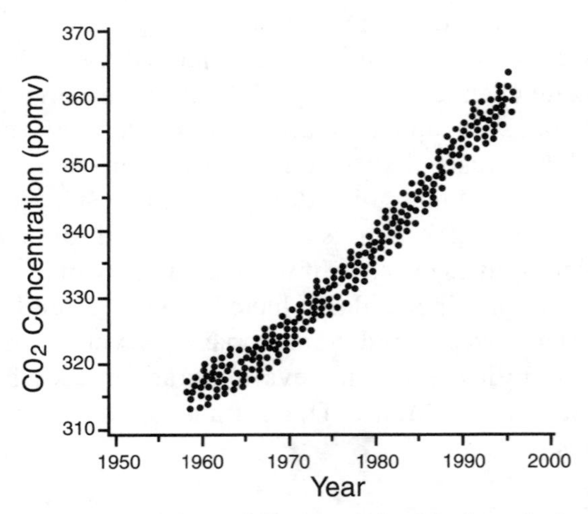

Figure 11.1. Rise of Carbon Dioxide Levels in the Atmosphere in the Last 40 Years.

From: C.D. Keeling and T.P. Whorf. 1997.

Why the Mojave Desert?

Theoretically, plant and ecosystem responses to elevated atmospheric CO_2 increase under conditions of reduced water availability. In fact, current predictions are that the productivity of desert ecosystems will increase by about 50%; this predicted increase for desert ecosystems is greater than for any other ecosystem type. The Mojave Desert is the driest ecosystem in North America and thus most likely to show a significant response to increased atmospheric CO_2. Since 40% of the earth's land is considered arid or semi-arid, even relatively small effects of CO_2 in a given model desert system, like the Mojave, can be compounded across deserts over the entire earth surface, resulting in potentially major global impacts. Furthermore, changes in how plants use water, coupled with the increased productivity of the desert under elevated CO_2, could have tremendous feedbacks on the hydrological cycle. Ultimately, these feedbacks may not only affect the function of ecological systems, but may also affect the availability of water to humans inhabiting arid and semi-arid regions. In the United States, it is these areas that are experiencing the fastest population growth.

The Nevada Desert FACE Facility is the first and only FACE facility in a desert ecosystem, and thus serves as the model for desert ecosystems around the world. Nevada Desert FACE is especially relevant to desertification, which is a major land use problem in North America as well as globally.

How Does FACE Technology Work?

A large fan next to an instrument shelter mixes ambient air with pure CO_2, and then forces that air-CO_2 mixture through a large pipe called the plenum (Figure 11.2). The plenum distributes the air-CO_2 mixture around the study plot via a circular ring of vent pipes. A computer regulates opening and closing of individual control valves based on the wind direction. The air-CO_2 mixture flows through upwind vent pipes to distribute the air-CO_2 mixture across the entire plot. The computer also monitors CO_2 concentration of air in the plot every second and updates the flow of pure CO_2 into the fan every second to maintain the plot's atmospheric CO_2 concentration at the target value of 550 ppm.

Why Place This Facility at the Nevada Test Site?

The NTS has a number of advantages, including a long research history that has concentrated on the structure and function of desert ecosystems and the environmental physiology of desert organisms. The entire area has been free from off-road recreational use or grazing by domestic livestock

Figure 11.2. FACE Apparatus

for more than the past 50 years; security at the NTS allows expensive, sophisticated instrumentation to be placed in the field on a continuous basis with no threats from theft or vandalism. Infrastructure at the NTS brings considerable logistical and technical support to a relatively remote desert site, as well as providing accommodations and meals for research staff and visiting scientists at the nearby facility of Mercury.

What Are the Scientific Studies Being Conducted at Desert FACE?

The Desert FACE studies test hypotheses that span levels of biological integration from cellular to ecosystem processes and are supported by experiments in controlled environment growth chambers and glasshouses as part of the Nevada Global Environmental Change Program (NevGEC). The primary objectives are:

Objective 1. Leaf- to Plant-Level Responses of Desert Vegetation to Elevated CO_2

Investigate the first and second order effects of increased atmospheric CO_2 concentration on the physiological processes of **photosynthesis**, allocation of carbon and other resources, shoot and root growth, and plant water relations in desert species with contrasting **growth form**, **phenology**, and photosynthetic pathways (Figure 11.3).

Figure 11.3. Researchers Acquiring Leaf-Level Photosynthesis Data with a Fluorometer and Gas Exchange System. A "sled" is used to sample within the FACE plots to prevent disturbance of the microbiotic crust.

Figure 11.4. Numerous measurements are collected to examine the effects of elevated atmospheric CO_2 on above and below ground plant tissue, soil, and insects. Depicted here are litter trays, nitrogen collars that help assess soil nitrogen cycling, and a soil respiration chamber. Other routine measurements include: soil moisture, air temperature, wind speed and direction, solar radiation, precipitation, plant photosynthesis and pigment content, leaf area, annual production, and perennial growth.

Objective 2. Ecosystem-Level Responses to Elevated CO_2

Determine the effect of elevated CO_2 on ecosystem-level processes such as carbon gain, **primary productivity**, water use and landscape water balance, nutrient uptake and cycling, and plant competition for resources that may result in community structure changes (Figures 11.4 and 11.5).

Objective 3. Build Predictive Models of Plant and Ecosystem Responses to Elevated CO_2

Integration between experimental ecology and modeling to simulate the temporal and spatial behavior of desert ecosystems and allow predictions of responses to a range of global changes including elevated CO_2 (Figure 11.6).

What Are Some of the Predicted Outcomes?

Increased atmospheric CO_2 levels will likely change plant water use patterns. Changes in plant water use may then have cascading effects on the hydrologic water balance, which in turn may affect the integrity of waste management programs, the

Figure 11.5. NDFF Ecologist collecting information on plant status in February 1999. The white tubes are protective covers for minirhizotron tubes; plexiglass tubes that provide access for a special video camera that is used to record root morphology and growth.

Yucca Mountain repository, and other projects of interest to the Nevada Test Site.

The effects of increased CO_2 levels will likely vary from species to species. These differential effects may cause changes in competitive balances among plant species, and therefore shifts in species abundance and changes in biodiversity and wildlife habitat. Of special interest is the exotic annual grass, red brome, whose predicted increased abundance may result in decreased water availability to native plants, as well as increased fire frequency and loss of biodiversity.

A potential beneficial effect is that increased growth and water use efficiency with increased CO_2 levels may enhance restoration efforts in desert ecosystems.

Nevada Desert FACE Facility – Results to Date

Some of the important results obtained after the experiment was initiated in April 1997 through the 1998 growing season are given below. Note that a very strong El Niño influenced the 1998 growing season with precipitation over twice the long-term average. Of particular interest is the response (percent change, adjusted by the density of individual plants) of total biomass, seed production, and seed quality to elevated CO_2 for several important plant species in the Mojave Desert. Our results to date validate the hypothesis, stated by several researchers, that desert regions may be the most responsive to elevated CO_2. The table below highlights these responses for: (1) *Bromus madritensis*, an invasive grass, (2) *Vulpia octoflora*, a native grass, and (3) *Lepidium lasiocarpum*, a native herb. All together, these plants accounted for a large number of the total cover of annual plants at NDFF in the 1998 growing season.

Plants generally increased in size and number at high CO_2, resulting in greater total biomass production per unit area, suggesting that elevated CO_2 increases the productivity of annual plants in desert ecosystems. This increase in plant growth also resulted in increases in the number of seeds produced per unit area. Seed "rain" (the production and transport of seeds) is important in the regeneration of future plant populations and as the major food source for many desert animals. The increase in seed rain may be offset somewhat by a general decrease in seed quality (as measured by seed nitrogen content). Species-specific differences in growth enhancement by elevated CO_2 resulted in qualitative differences in these performance characteristics. For example, the invasive grass was much more responsive with regard to growth and reproduction than the native grass in this growing season. How the quantitative and qualitative changes in biomass and seed production at elevated CO_2 may impact plant and animal diversity in the Mojave Desert is the subject of current research.

Figure 11.6. This photograph was acquired Spring 1998 after an El Niño. The lush green of the annual and perennial herbaceous cover clearly shows the effect of increased rainfall and seasonal changes as compared to the above winter photograph taken in a La Niña (dry) year.

Percent Change at Elevated CO_2

Species	Biomass	Seed Rain	Seed Quality
Bromus madritensis	+92 %	+210 %	- 10 %
Lepidium lasiocarpum	+24 %	+43 %	- 13 %
Vulpia octoflora	0 %	- 25 %	- 9 %
C. Total	+55 %	+46 %	

Changes in PrimaryProduction

- Photosynthesis of all perennial and annual C_3 species that have been studied was enhanced by elevated CO_2 by up to 50%. No effects have occurred in a C_4 grass.
- Similar enhancements in primary production also occurred:

  Increased leaf area index of *Larrea*, supported by spectral reflectance and tagged shoot measurements.

  Increased annual plant production.
- Photosynthetic down-regulation occurred in some but not all C_3 species and depended on water availability. For *Larrea* and *Krameria*, down-regulation is pronounced in wet seasons, but not in dry seasons. Elevated CO_2 also reduced stress in photosystem II (PSII) by serving as an enhanced sink for energy flow through PSII.
- Leaf-level water use efficiency increased under elevated CO_2 during the wet season.

Changes in Plant Reproduction and Recruitment

- A trade-off between growth and reproduction occurs in *Bromus* when grown at elevated CO_2. Seeds from parents grown at elevated CO_2 produced plants with lowered growth potential.
- Seedling survival for the two dominant shrubs, *Larrea* and *Ambrosia*, were significantly higher under elevated CO_2.

Changes in Water Balance

- Gas exchange measurements indicate a general pattern of reduced **stomatal conductance** under elevated CO_2 during the wet season, but no effect during the dry season. The extent and timing of the reduction varied among growth forms.
- Sap flow data indicate reduced canopy-level transpiration for the leafless shrub *Ephedra*, but not for the evergreen shrub *Larrea*. However, both species had decreased ratios of transpiration area to sapwood area. Thus, plant hydraulics improved with elevated CO_2, and in contrast to the leaf level measurements, canopy stomatal conductance was either unchanged or *increased* under elevated CO_2.
- No differences in leaf temperatures were found for the **microphyllous** shrub *Larrea*.
- Soil moisture for the top 0.5 m of soil indicate no differences in soil moisture in shrub interspaces, but a small, transient increase under elevated CO_2 beneath shrub canopies.

Changes in Nitrogen Cycling

- The microbiotic crust at the NDFF is capable of **nitrogen fixation**, although species composition of the crust influences the rate.
- Nitrogen fixation also occurs, and its rate may exceed that of the crust.
- Nitrogen fixation, potential denitrification, and soil ammonium content were greater under *Larrea* canopies than in the plant interspace.

Changes in Belowground Processes

- Soil respiration integrated over a 24-hour period during the dry season tended to be greater under elevated CO_2.
- During the winter, the evergreen shrub *Larrea* produced significantly greater fine root length in the top 0.5 m of soil under elevated CO_2. However by late spring, root length was not significantly different between CO_2 treatments.
- In a glasshouse study of a C_3 annual, a C_3 perennial, and a C_4 perennial grass, root biomass was only increased in the C_3 perennial grass. Little evidence for up-regulation in root physiology or nutrient uptake in response to CO_2 was observed in any of the three grasses.

Mojave Global Change Facility (MGCF)—A Related Research Effort

A new effort has been initiated to examine the impact of global climate change factors other than increased CO_2 on the Mojave Desert Ecosystem. How CO_2 is likely to alter a desert ecosystem is driven by two key factors: water and nitrogen availability. CO_2 is likely to alter a desert ecosystem by changing the availability of water, which in turn may regulate plant responses to CO_2. But the degree to which that system responds to elevated CO_2 will be driven by the availability of critical nutrients, particularly nitrogen. Global climate change models have predicted that rising levels of atmospheric CO_2 will increase summer monsoon rains in the Mojave Desert. Predicted increases in urbanization and cattle grazing will result in increased nitrogen deposition and disturbance or destruction of the soil crust. The **biological soil crust**, which is present at the surface of most desert soils, plays an important role in making nitrogen available to desert plants. Thus, the explicit examination of how changing water and nitrogen availability affect a desert ecosystem, when combined with our knowledge of how CO_2 affects a desert, will allow us to form the most comprehensive understanding possible for how arid ecosystems will respond to global change.

What Experimental Treatments Are Being Applied?

Three manipulations based on these predictions are being applied to the 96 MGCF plots (each 14 by 14 m). These treatments include three summer irrigation treatments (25 mm each), nitrogen fertilization (10 and 40 kg/ha), and crust disturbance in factorial combination resulting in a total of eight treatment combinations: (1) control; (2) plus summer rainfall; (3) plus crust disturbance; (4) plus nitrogen deposition; (5) plus summer rainfall and crust disturbance; (6) plus summer rainfall and nitrogen deposition; (7) plus crust disturbance and nitrogen deposition; and (8) plus all three treatments. A randomized block design was used to establish the treatments for each plot.

Although the seventh treatment combination, plus crust disturbance and nitrogen deposition, may seem unusual in that one treatment reduces nitrogen availability whereas the other increases N availability, this treatment combination is very realistic. This is because much of the desert Southwest has experienced significant losses of the biological soil surface crust due to land use change (e.g., livestock grazing and recreational activities) while simultaneously experiencing increasing rates of nitrogen deposition due to rapidly increasing population growth (and therefore urban air pollution).

Control plots will be undisturbed, unmanipulated plots in their natural condition. The monitoring of these plots over the long term will provide detailed information on the effects of urbanization (for example, growth of Las Vegas) on a nearly pristine ecosystem.

The results from the MGCF will be combined with data from the NDFF to make predictions on the overall impact of future climate change on the Mojave Desert, which in turn can assist with better land management, restoration, revegetation, and various clean-up efforts (Figure 11.7)

What Are Some of the Predicted Outcomes?

- Increased nitrogen deposition will result in increases in plant photosynthesis and primary production, particularly in concert with increased rainfall.
- Disturbance of biological soil surface crusts will reduce N-inputs into the soil, which will

Figure 11.7. Building the Mojave Global Change Facility progressed over one year. 96 plots were laid out, over 1000 posts were installed, and a weather station was assembled to collect meteorological data. The 40-foot crane was used to collect images of each plot to establish baseline information.

result in N-depletion in both soils and vegetation and in reduced plant production over time.

- Increased summer rainfall will result in increased production; growth forms such as summer annuals and perennial grasses will increase more in production than drought-tolerant **xerophytic** shrubs.
- Changes in the rate of ecosystem water use will be proportional to changes in productivity, but ecosystem water balance will not be affected by the treatments because plants will fully extract all available soil water.
- Both increased N-deposition and surface crust disturbance will stimulate growth of an exotic annual grass to a greater extent than for native annual species.

What Studies Are Being Initiated to Measure Predicted Outcomes?

- Water uptake by roots
- Soil heterogeneity
- Nitrogen fixation
- Moss responses
- Soil carbon uptake and respiration
- Leaf/canopy level photosynthesis
- Primary productivity
- Soil moisture and temperature differences
- Reflectance measurements
- Aerial photography

How Can the Studies Outlined Above be Used to Test the Predicted Outcomes?

From the discussions in this chapter, you should be able to understand the rationale behind most of the measurements being made. Now, connect those measurements with the hypotheses, the variables, and the predictions.

Have Any Differences between Treatments Been Observed?

Results to date are qualitative, but as anticipated, increases in biomass and photosynthetic

Figure 11.8. Photographs of *Larrea tridentata* taken in irrigated plot (left) and non-irrigated plot (right) after a below-average winter precipitation year.

capacity are occurring during peak heat and drought (Figure 11.8).

Selected Publications

Evans RD, Johansen, JR (1999) Microbiotic crusts and ecosystem processes. *Critical Reviews in Plant Sciences* 18:183-225.

Hamerlynck EP, Huxman TE, Loik ME, Smith SD (2000) Effects of extreme high temperature, drought and elevated CO_2 on photosynthesis of *Larrea tridentata*, a Mojave Desert evergreen. *Plant Ecology* 148:185-195.

Huxman TE, Hamerlynck EP, Smith SD (1999) Reproductive allocation and seed production in *Bromus madritensis* ssp. *rubens* at elevated atmospheric CO_2. *Functional Ecology* 13:769-777.

Jordan DE, Zitzer SF, Hendrey GR, Lewin KF, Nagy J, Nowak RS, Smith SD, Coleman JS, Seemann JR (1999) Biotic, abiotic and performance aspects of the Nevada Desert Free-Air CO_2 enrichment (FACE) facility. *Global Change Biology* 5:659-668.

Nowak RS, Jordan DN, DeFalco LA, Wilcox CS, Coleman JS, Seemann JR, Smith SD (2001) Leaf conductance decreased under Free-air CO_2 Enrichment (FACE) for three perennials in the Nevada desert. *New Phytologist* 150:449-458.

Pataki DE, Huxman TE, Jordan DN, Zitzer SF, Coleman JS, Smith SD, Nowak RS, Seemann JR (2000) Water use of Mojave Desert shrubs under elevated CO_2. *Global Change Biology* 6:889-898.

Smith SD, Huxman TE, Zitzer SF, Charlet TN, Housman DC, Coleman JS, Fenstermaker LK, Seemann JR, Nowak RS (2000) Elevated CO_2 increases productivity and invasive species success in an arid ecosystem. *Nature* 408:79-82.

Yoder CK, Vivin P, DeFalco L, Seemann JR, Nowak RS (2000) Root function in three Mojave Desert grasses in response to atmospheric CO_2 enrichment and soil water stress. *New Phytologist* 145:245-256.

Definitions

annual: A plant that germinates from seed, flowers, produces seed, and dies within one year.

biodiversity: The number and variety of organisms found within a specified geographic region.

biological soil crust: Living surfaces present on desert soils dominated by blue-green algae, lichens, mosses, green algae, microfungi, and bacteria. Biological soil crusts are important components of nutrient cycling and soil stabilization in arid and semiarid ecosystems throughout the Southwest.

carbon cycle: In ecology, the movement of atoms of carbon through the biosphere. Molecules of carbon dioxide are taken in by plants, to be incorporated into their tissues, which may then be eaten by and incorporated into animals. Animals return the carbon to the air in the form of carbon dioxide, and the cycle starts again.

carbon dioxide: Carbon dioxide is normally found as a gas that is breathed out by animals during respiration and absorbed by green plants during photosynthesis. It is part of the atmosphere, making up less than 1% of the volume of dry air. It is sometimes referred to a "greenhouse gas" because it traps heat from solar radiation and raises the temperature of the atmosphere.

deciduous: A growth form characterized by the periodic loss of leaves in response to a particular physiological or environmental signal, for example, drought or winter cold.

denitrification: The chemical reduction of nitrates (NO_3^-) or nitrites (NO_2^-) to gas (N_2), by bacterial action in soil.

desertification: The transformation of arable or habitable land to desert by a change in climate or destructive land use.

ecotone: A transitional zone between two communities containing the characteristic species of each.

evergreen: A growth form characterized by retention of green leaves throughout the year.

growth form: Equivalent to "life form" as used in Chapter 3 (Nevada Biomes). The form that primarily defines a plant as a tree, shrub, or herb. May be deciduous or evergreen, annual, perennial, or ephemeral. Further characteristics may be defined as needed by context (e.g., leaf traits (broadleaf, needle-leaf, etc.), independence (parasitic, saprophytic, etc.), or seasonality.

hydrologic cycle: The continuous circular process in which the water of the earth evaporates from the oceans, condenses, falls to the earth as rain or snow, and eventually returns to the oceans through runoff in rivers or streams. Some water is absorbed by plants and returned to the atmosphere as vapor.

microphyllous: An adjective used to describe a plant with small leaves. Many desert plants are microphyllous.

nitrogen fixation: The conversion by certain soil microorganisms, such as rhizobia, of atmospheric nitrogen (N_2) into NH_4^+, which plants and other organisms can assimilate.

perennial: Any plant that under natural conditions lives for several to many growing seasons, as contrasted to an annual or a biennial plant. Botanically, the term *perennial* applies to both woody and herbaceous plants.

phenology: The relationship between periodic biological phenomena (e.g., flowering, breeding, and migration) and climatic conditions.

photosynthesis: The process in green plants and certain other organisms by which carbohydrates are synthesized from carbon dioxide and water using light as an energy source. Most forms of photosynthesis release oxygen as a byproduct.

primary productivity: The net accumulation of living tissue in a plant as a result of photosynthesis.

physiognomy: The study and classification of the physical characteristics of vegetation.

reflectance: The fraction of the total radiant flux incident upon a surface that is reflected and that varies according to the wavelength distribution of the incident radiation and also the characteristics of the leaf surface.

respiration: The overall process by which oxygen is extracted from air and is transported to the cells for the oxidation of organic molecules while carbon dioxide (CO_2) and water, the products of oxidation, are returned to the environment.

stomatal conductance: A measure of the relative opening of plant pores (stomates) in the epidermis of a leaf or stem. This occurs as an inevitable consequence of photosynthesis with the uptake of carbon dioxide, and results in plant water loss (transpiration).

surface albedo: The fraction of incident electromagnetic radiation reflected by a surface. In the visible spectrum, light surfaces reflect more than dark; smooth surfaces reflect more than rough.

transpiration: The loss of water by evaporation from terrestrial plants. Some evaporation occurs directly through the exposed walls of surface cells, but the greatest amount takes place through the stomates. Transpiration causes the ascent of sap from the roots to the leaves (thus supplying the water needed for photosynthesis) and provides the moisture necessary for the diffusion of carbon dioxide and oxygen.

xerophytic: Pertaining to a plant adapted to living in a dry arid habitat; a desert plant (xerophyte).

General References

Alcorn, J. R. 1988. The birds of Nevada. Fairview West Publishing, Fallon, Nevada. 418 pp.

Anderson, B. W., and K. E. Holte. 1981. Vegetation development over 25 years without grazing on sagebrush-dominated rangeland in southeastern Idaho. *Journal of Range Management* 34:25-29.

Austin, G.T. 1985. Lowland riparian butterflies of the Great Basin and associated areas. The *Journal of Research on the Lepidoptera* 24:117-131.

Axelrod, D. I. 1950. Evolution of desert vegetation in Western North America. Contributions to Paleontology. Carnegie Institute of Washington Publication, 590 VI. 306 pp.

Axelrod, D. I. 1976. History of the coniferous forests, California and Nevada. University of California Publications in Botany 70. 62 pp.

Axelrod, D. I. 1983. Paleobotanical history of the Western Deserts. Pages 113-129 *in* S.G. Wells and D.R. Haragan (*editors*). Origin and evolution of deserts. University of New Mexico Press, Albuquerque.

Axelrod, D. I. and P. H. Raven. 1985. Origins of the Cordilleran flora. *Journal of Biogeography* 12:21-47.

d'Azavedo, W. 1986. Handbook of North American Indians, Volume 11: Great Basin. Smithsonian Institution, Washington, D.C. 852 pp.

Barrows, C. W. 1993. Tamarisk control II: a success study. *Restoration and Management Notes* 11:35-38.

Beatley, J. 1966. Ecological status of introduced brome grasses (*Bromus* spp.) in desert vegetation of southern Nevada. *Ecology* 47(4):548-554.

Beatley, J. C. 1974. Phenological events and their environmental triggers in Mojave Desert ecosystems. *Ecology* 55:856-863.

Beatley, J. C. 1974. Effects of rainfall and temperature on distribution and behavior of *Larrea tridentata* (creosote-bush) in the Mojave Desert of Nevada. *Ecology* 55:245-261.

Beatley, J. C. 1976. Vascular plants of the Nevada Test Site and central-southern Nevada: ecologic and geographic distributions, TID-26881. Energy Research and Development Administration, National Technical Information Service, Springfield, Virginia.

Behle, W. H. 1978. Avian biogeography of the Great Basin and Intermountain Region. Intermountain biogeography: a symposium. *Great Basin Naturalist Memoirs* 2:55-80.

Belovsky, G. E. 1987. Extinction models and mammalian persistence. Pages 35-58 *in* M. E. Soulé (*editor*). Viable populations for conservation, Cambridge University Press, Cambridge, England.

Benson, L. V., D. R. Currey, R. I. Dorn, K. R. Lajoie, C. G. Oviatt, S. W. Robinson, G. I. Smith, and S. Stine. 1990. Chronology of expansion and contraction of four Great Basin lake systems during the past 35,000 years. *Palaeogeography, Palaeoclimatology, Palaeoecology* 78:241-286.

Betancourt, J. L., T. R. Van Devender and P. S. Martin. 1990. Introduction. Pages 2-11 *in* J. L. Betancourt, T. R. Van Devender and P. S. Martin (*editors*). Packrat middens: 40,000 years of biotic change. The University of Arizona Press, Tucson.

Bettinger, R. L. 1991. Native land use: Archaeology and anthropology. Pages 463-486 *in* C. A. Hall, Jr. (*editor*), Natural History of the White-Inyo Range, eastern California. University of California Press, Berkeley and Los Angeles.

Billings, W. D. 1949. The shadscale vegetation zone of Nevada and eastern California in relation to climate and soils. *American Midland Naturalist* 42:87-109.

Billings, W. D. 1954. Nevada trees. Agricultural Extension Service, Bulletin 94, University of Nevada, Reno. 125 pp.

Billings, W. D. 1970. Plants and the ecosystem. Wadsworth Publishing Company, Inc., Belmont, California. iv + 154 pp.

Billings, W. D. 1978. Alpine phytogeography across the Great Basin. Intermountain biogeography: a symposium. *Great Basin Naturalist Memoirs* 2:105-117.

Billings, W. D. 1990. The mountain forests of North America and their environments. Pages 47-86 *in* C.B. Osmond, L.F. Pitelka and G.M. Hidy (*editors*). Plant biology of the Basin and Range. Springer-Verlag, New York.

Billings, W. D. and A. F. Mark. 1957. Factors involved in the persistence of montane treeless balds. *Ecology* 38:140-142.

Blackburn, W. H. and P. Tueller. 1970. Pinyon and juniper invasion in black sagebrush communities in east-central Nevada. *Ecology* 51:841-848.

Blackwelder, E. 1931. Pleistocene glaciation in the Sierra Nevada and Basin ranges. *Geological Society of America Bulletin* 42:865-922.

Blackwelder, E. 1934. Supplementary notes on Pleistocene glaciation in the Great Basin. *Journal of the Washington Academy of Sciences* 24:212-222.

Bock, C. E., V. A. Saab, T. D. Rich, and D. S. Dobkin. 1993. Effects of livestock grazing on neotropical migratory landbirds in western North America. Pages 296-309 *in* D. M. Finch and P. W. Stangel (editors). Status and management of neotropical migratory birds. U.S.D.A. Forest Service General Technical Report RM- 229.

Bohning-Gaese, K., M. L. Taper, and J. H. Brown. 1993. Are declines in North American insectivorous songbirds due to causes on the breeding range? *Conservation Biology* 7:76-86.

Brown, J. H. 1971. Mammals on mountaintops: nonequilibrium insular biogeography. *American Naturalist* 105:467-478.

Brown, J. H. 1973. Species diversity of seed-eating rodents in sand dune habitats. *Ecology* 54:775-787.

Brown, J. H. 1978. The theory of insular biogeography and the distribution of boreal birds and mammals. *Great Basin Naturalist Memoirs* 2:209-227.

Brussard, P. F., and G. A. Austin. 1993. Nevada butterflies: Check list and ecological distribution. Nevada Biodiversity Initiative, Reno. 5 pp.

Brussard, P.F., D.A. Charlet, and D. Dobkin. 1999. The Great Basin-Mojave Desert Region. Pages 505-542 *in*: Mac, M.J., P.A. Opler, C.E. Puckett-Haecker, and P.D. Doran (*editors*). The status and trends of the nation's biological resources. U.S. Department of the Interior, U.S. Geological Survey, Reston, Virginia.

Bunting, S. C. 1986. Use of prescribed burning in juniper and pinyon-juniper woodlands. Pages 141-144 *in*: R. L. Everett (*editor*), Proceedings: Pinyon-juniper conference. U.S. Forest Service General Technical Report INT-215.

Burt, W. H., and R. P. Grossenheider. 1976. Mammals. Peterson Field Guides. Houghton Mifflin Company. Boston, Mass. xvii + 289 pp.

Carothers, S. W., R. R. Johnson, and S. W. Aitchison. 1974. Population structure and social organization of Southwestern riparian birds. *American Zoologist* 14:97-108.

Carter, M. F., and K. Barker. 1993. An interactive database for setting conservation priorities for western neotropical migrants. Pages 120-144 *in* D. M. Finch and P. W. Stangel (*editors*). Status and management of neotropical migratory birds. U.S.D.A. Forest Service General Technical Report RM-229.

Chandler, J. R. 1970. A biological approach to water quality management. *Water Pollution Control* 4:415-22.

Chaney, E., W. Elmore, and W.S. Platts. 1993. Livestock grazing on western riparian areas. US Government Printing Office, Washington, D.C. 45 pp.

Chapman, H. C. 1966. The mosquitoes of Nevada. Entomology Research Division, Agriculture Research Service, U.S. Department of Agriculture and the Max C. Fleischmann College of Agriculture, University of Nevada. Reno. 41 pp.

Charlet, D. A. 1991. Relationships of the Great Basin alpine flora: A quantitative analysis. M.S. Thesis, Department of Biology, University of Nevada, Reno.

Charlet, D. A. 1996. Atlas of Nevada conifers: A phytogeographic reference. University of Nevada Press, Reno. 336 pp.

Charlet, D. A., and R. W. Rust. 1991. Visitation of high mountain bogs by Golden Eagles in the northern Great Basin. *Journal of Field Ornithology* 62:46-52.

Clark County. 2000. Clark County Multiple Species Habitat Conservation Plan and Environmental Impact Statement for Issuance of a Permit to Allow Incidental Take of 79 Species in Clark County, Nevada. Clark County Department of Comprehensive Planning. Prepared by RECON, San Diego, California.

Clark, D. R., Jr. and R. L. Hothem. 1991. Mammal mortality at Arizona, California, and Nevada gold mines using cyanide extraction. *California Fish and Game* 77(2):61-69.

Critchfield, W. B. 1984b. Impact of the Pleistocene on the genetic structure of North American conifers. Pages 70-118 *in* R. L. Lanner (*editor*). Proceedings of the eighth North American Forest Biology Workshop, July 30 - August 1, 1984. Utah State University, Logan, Utah.

Critchfield, W. B. and G. L. Allenbaugh. 1969. The distribution of Pinaceae in and near northern Nevada. *Madroño* 20:12-26.

Crum, S.J. 1994. The road on which we came: A history of the Western Shoshone. University of Utah Press, Salt Lake City. 240 pp.

D'Antonio, C. M. and P. M. Vitousek. 1992. Biological invasions by exotic grasses, the grass/fire cycle, and global change. *Annual Review of Ecology and Systematics* 23:63-87.

Deacon, J. E. 1979. Endangered and threatened fishes of the west. *Great Basin Naturalist Memoirs* 3:41-64.

Deacon, J. E. and C. D. Williams. 1991. Ash Meadows and the legacy of the Devils Hole pupfish. Pages 69-87 *in* W. L. Minckley and J. E. Deacon (*editors*). Battle against extinction. The University of Arizona Press, Tucson.

DeLucia, E. H., and W. H. Schlesinger. 1990. Ecophysiology of Great Basin and Sierra Nevada vegetation on contrasting soils. Pages 143-178 *in* C.B. Osmond, L. F. Pitelka and G. M. Hidy (*editors*). Plant biology of the Basin and Range. Springer-Verlag, Berlin, Germany.

DeLucia, E. H., W. H. Schlesinger, and W. D. Billings. 1988. Water relations and the maintenance of Sierran conifers on hydrothermally altered rock. *Ecology* 69:303-311.

Demarais, B. D., T. E. Dowling and W. L. Minckley. 1993. Post-perturbation genetic changes in populations of endangered Virgin River chubs. *Conservation Biology* 7:334-341.

Desante, D. F., and T. L. George. 1994. Population trends in the landbirds of western North America. Pages 173-190 *in* J. R. Jehl, Jr., and N. K. Johnson (*editors*). A century of avifaunal change in western North America. Studies in Avian Biology, Volume 15.

Desert Research Institute. 1968. Final Reports on the Lehman Caves studies to the Department of Interior, The U. S. National Park Service, Lehman Caves National Monument. The Laboratory of Desert Biology, Desert Research Institute, Reno, Nevada. 57 pp.

Dobkin, D. S. 1994. Community composition and habitat affinities of riparian birds on the Sheldon-Hart Mountain Refuges, Nevada and Oregon, 1991-93. Final Report. U.S. Fish and Wildlife Service, Lakeview, Oregon. 287 pp.

Dobkin, D. S. 1996. Conservation and management of neotropical migrant landbirds in the Great Basin. University Idaho Press, Moscow, Idaho.

Dobkin, D. S. and B. A. Wilcox. 1986. Analysis of natural forest fragments: riparian birds in the Toiyabe Mountains, Nevada. Pages 293-299 *in* J. Verner, M. L. Morrison, and C. J. Ralph (*editors*). Wildlife 2000: modeling habitat relationships of terrestrial vertebrates. University of Wisconsin Press, Madison, Wisconsin.

Dohrenwend, J. C., W. B. Bull, L. D. McFadden, G. I. Smith, R. S. U. Smith, and S. G. Wells. 1991. Quaternary geology of the Great Basin. Pages 321-352 *in* R. B. Morrison (*editor*). Quaternary nonglacial geology: Conterminous U.S., vol. K-2, Geological Society of America, Boulder, Colorado.

Ehrlich, P. R., and D. D. Murphy. 1987. Monitoring populations on remnants of native vegetation. Pages 201-210 *in* D. A. Saunders, G. W. Arnold, A. A. Burbidge, and A. J. M. Hopkins (*editors*). Nature conservation: the role of remnants of native vegetation. Surrey Beatty and Sons Pty Limited, Chipping Norton, New South Wales, Australia.

El-Ghonemy, A. A., A. Wallace, and E. M. Romney. 1980. Socioecological and soil-plant studies of the natural vegetation in the northern Mojave Desert-Great Basin desert interface. *Great Basin Naturalist Memoirs* 4:71-86.

Elliot-Fisk, D.L., T.C. Cahill, O.K. Davis, L. Duan, C.R. Goldman, G.E. Gruell, R. Harris, R. Kattelmann, R. Lacey, D. Leisz, S. Lindstrom, D. Machida, R.A. Rowntree, P. Rucks, D.A. Sharkey, S. Stephens, and D.S. Ziegler. 1997. Lake Tahoe Case Study. Sierra Nevada Ecosystem Project: Final Report to Congress, Addendum. University of California, Davis, Centers for Water and Wildland Resources.

Fiero, B. 1986. Geology of the Great Basin. Max C. Fleischmann series in Great Basin natural history. University of Nevada Press, Reno. 197 pp.

Fleischner, T. L. 1994. Ecological costs of livestock grazing in western North America. *Conservation Biology* 8:629-644.

Flint, R. F. 1971. Glacial and Quaternary geology. John Wiley and Sons, Inc., New York. 892 pp.

Frémont, J. C. 1845. Report of the Exploring Expedition to the Rocky Mountains in the year 1842, and to Oregon and California in the Years 1843-1844. Goles and Seaton, Washington, D.C. 693 pp.

Goodwin, J. 1992. The role of mycorrhizal fungi in competitive interactions among native bunchgrasses and alien weeds: a review and synthesis. *Northwest Science* 66:251-260.

Gould, S. J. 1991. Abolish the recent. *Natural History* 5/91:16-21.

Graf, W. L. 1980. Riparian management: a flood control perspective. *Journal of Soil and Water Conservation* 35:158-161.

Grayson, D. K. 1987. The biogeographic history of small mammals in the Great Basin: observation on the last 20,000 years. *Journal of Mammology* 68:359-375.

Grayson, D. K. 1993. The desert's past: A natural prehistory of the Great Basin. Smithsonian Institution Press, Washington, D.C. 356 pp.

Hafner, D. J. 1992. Speciation and persistence of a contact zone in Mojave Desert ground squirrel, subgenus *Xerospermophilus*. *Journal of Mammology* 73:770-778.

Hall, E. R. 1946. Mammals of Nevada. University of California Press, Berkeley, California. xi + 710 pp.

Hammond, P. C., and D. V. McCorkle. 1983. The decline and extinction of *Speyeria* populations resulting from human environmental disturbances (Nymphalidae: Argynninae). *The Journal of Research on the Lepidoptera* 22:217-224.

Harper, K. T., D. L. Freeman, W. K. Ostler, and L. G. Klikoff. 1978. The flora of Great Basin mountain ranges: diversity, sources, and dispersal ecology. Pages 81-104 *in* K. T. Harper and J. L. Reveal (*editors*). Intermountain biogeography: A symposium. *Great Basin Naturalist Memoirs* 2.

Hershler, R. and D. Sada. 1987. Springsnails (Gastropoda; Hydrobiidae) of Ash Meadows, Amargosa Basin, California-Nevada. *Proeeedings of the Biological Society of Washington* 100:776-843.

Hidy, G. M. and H. E. Klieforth. 1990. Atmospheric processes affecting the climate of the Great Basin. Pages 17-45 *in* C. B. Osmond, L. F. Pitelka, and G. M. Hidy (*editors*). Plant biology of the Basin and Range. Springer-Verlag, New York.

Holmgren, N. 1972. Plant geography in the Intermountain Region. Pages 77-161 *in* A. Cronquist, A. H. Holmgren, N. H. Holmgren, and J. L. Reveal (*editors*). Intermountain flora. Volume I. Hafner Publishing Company, New York.

Horne, A. J., J. C. Roth, and N. J. Barratt. 1994. Walker Lake, Nevada: State of the lake 1992-1994. Report to the Nevada Department of Environmental Protection. 83 pp. + appendices.

Houghton, J. G., C. M. Sakamoto, and R. O. Gifford. 1975. Nevada's weather and climate. Nevada Bureau of Mines and Geology, Special Publication 2, Reno. 78 pp.

Hunt, C. B. 1967. Physiography of the United States. W. H. Freeman, San Francisco, California. 480 pp.

Jehl, J. R., Jr. 1994. Changes in saline and alkaline lake avifaunas in western North America in the past 150 years. Pages 258-272 *in* J. R. Jehl, Jr., and N. K. Johnson (*editors*). A century of avifaunal change in western North America. Studies in Avian Biology, Volume 15.

Johnson, N. K. 1975. Controls of number of bird species on montane islands in the Great Basin. *Evolution* 29:545-567.

Johnson, N. K. 1978. Patterns of avian geography and speciation in the Great Basin. Pages 137-159 *in* K. T. Harper and J. L. Reveal (*editors*). Intermountain biogeography: a symposium. *Great Basin Naturalist Memoirs* 2.

Johnson, R. R., C. D. Ziebell, D. R. Patton, P. F. Folliott, and R. H. Hamre (*technical editors*). 1985. Riparian ecosystems and their management: Reconciling conflicting uses. U.S. Forest Service General Technical Report RM-120. 523 pp.

Johnson, T. N. 1986. Using herbicides for pinyon-juniper control in the southwest. Pages 330-334 *in* R. L. Everett (*editor*), Proceedings: Pinyon-juniper conference. U.S. Forest Service General Technical Report INT-215.

Kartesz, J. T. 1988. A flora of Nevada. Ph.D. dissertation, Department of Biology, University of Nevada, Reno. 1,729 pp.

Kauffman, J. B., and W. C. Krueger. 1984. Livestock impacts on riparian ecosystems and stream management implications: a review. *Journal of Range Management* 37:430-438.

King, P. B. 1977. The Evolution of North America. Princeton University Press, Princeton, New Jersey. 197 pp.

Knopf, F. L., R. R. Johnson, T. Rich, F. B. Samson, and R. C. Szaro. 1988. Conservation of riparian ecosystems in the United States. *Wilson Bulletin* 100:272-284.

Knopf, F. L., J. A. Sedgwick, and R. W. Cannon. 1988. Guild structure of a riparian avifauna relative to seasonal cattle grazing. *Journal of Wildlife Management* 52:280-290.

Kremen, C. 1992. Assessing the indicator properties of species assemblages for natural areas monitoring. *Ecological Applications* 2:203-217.

Kremen, C., R. K. Colwell, T. L. Erwin, D. D. Murphy, R. F. Noss, and M. A. Sanjayan. 1993. Terrestrial arthropod assemblages: Their use in conservation planning. *Conservation Biology* 7:796-808.

Krueper, D. J. 1993. Effects of land use practices on western riparian ecosystems. Pages 321-330 *in* D. M. Finch and P. W. Stangel (*editors*). Status and management of neotropical migratory birds. U. S. Forest Service General Technical Report RM-229.

Lancaster, N. 1988. Controls of eolian dune size and spacing. *Geology* 16:972-975.

Lancaster, N. 1988. On desert sand seas. *Episodes* 11:12-17.

Lancaster, N. 1989. The dynamics of star dunes: An example from Grand Desierto, Mexico. *Sedimentology* 36:273-289.

Lanner, R. M. 1984. Trees of the Great Basin. University of Nevada Press, Reno. 215 pp.

Las Vegas Valley Water District. 1992. Environmental Report of the Virgin River Water Resource Development Project, Clark County, NV. Cooperative Water Project, Report No. 2, Hydrographic Basin 222. 130 pp.

Las Vegas Valley Water District 1993. Addendum to Environmental Report of the Virgin River Water Resource Development Project, Clark County, NV. Cooperative Water Project, Report No. 2, Hydrographic Basin 222. 27 pp.

Lavin, M. T. 1981. The floristics of the headwaters of the Walker River, California and Nevada. M.S. thesis, Department of Biology, University of Nevada, Reno. 141 pp.

Linsdale, J. M. 1936. The birds of Nevada. Pacific coast avifauna No. 23, Cooper Ornithological Society, Berkeley, California. 145 pp.

Linsdale, J. M. 1938. Environmental responses of vertebrates in the Great Basin. American *Midland Naturalist* 19:1-206.

Linsdale, M. A., J. T. Howell, and J. M. Linsdale. 1952. Plants of the Toiyabe Mountains area, Nevada. *Wasmann Journal of Biology* 10:129-200.

Little, E. L., Jr. 1971. Atlas of United States trees. Volume 1. Conifers and important hardwoods. United States Department of Agriculture Forest Service. Miscellaneous Publication Number 1146. 203 pp.

Loope, L. L. 1969. Subalpine and alpine vegetation of northeastern Nevada. Ph.D. dissertation, Department of Botany, Duke University, Durham, North Carolina. 292 pp.

Lovich, J. E., T. B. Egan, and R. C. de Gouvenain. 1994. Tamarisk control on public lands in the desert of southern California: Two case studies. 46th Annual California Weed Conference, California Weed Science Society. 166-177 pp.

Luchenbach, R. A., and R. B. Bury. 1983. Effects of off-road vehicles on the biota of Algodones Dunes, Imperial County, California. *Journal of Applied Ecology* 20:265-286.

MacMahon, J. A. 1988. Warm deserts. Pages 231-264 *in* M. G. Barbour and W. D. Billings (*editors*). North American terrestrial vegetation. Cambridge University Press, New York.

MacMahon, J. A. 1988. North American deserts: their floral and faunal components. Pages 21-81 *in* M. G. Barbour and W. D. Billings (*editors*). North American terrestrial vegetation. Cambridge University Press, New York.

Mack, R. N. 1986. Alien plant invasion into the Intermountain West: A case history. Pages 191-213 *in* Mooney, H. A., and J. A. Drake, editors. Ecology of biological invasions in North America and Hawaii. Springer-Verlag, New York.

McLane, A. R. 1978. *Silent Cordilleras*. Camp Nevada, Reno. 118 pp.

Melgoza, G., R. S. Nowak, and R. J. Tausch. 1990. Soil water exploitation after fire: Competition between *Bromus tectorum* (cheatgrass) and two native species. *Oecologia* 83:7-13.

Meyer, S. E. 1978. Some factors governing plant distributions in the Mojave-Intermountain transition zone. Intermountain biogeography: A symposium. *Great Basin Naturalist Memoirs* 2:197-207.

Meyer, S. E. 1986. The ecology of gypsum communities in the Mojave Desert. *Ecology* 67:1303-1313.

Mifflin, M. D. and M. M. Wheat. 1979. Pluvial lakes and estimated pluvial climates of Nevada. Mackay School of Mines Bulletin Number 94. University of Nevada, Reno. 57 pp.

Miller, R. F., and P. E. Wigand. 1994. Holocene changes in semiarid pinyon-juniper woodlands. *BioScience* 44:465-474.

Minckley, W. L. and M. E. Douglas. 1991. Discovery and extinction of western fishes: a blink of the eye in geologic time. Pages 7-17 *in* W. L. Minckley and J.E. Deacon (*editors*). Battle against extinction. The University of Arizona Press, Tucson.

Morefield, J. D. 1992. Spatial and ecologic segregation of phytogeographic elements in the White Mountains of California and Nevada. *Journal of Biogeography* 19:33-50.

Morefield, J. D., D. W. Taylor and M. DeDecker. 1988. Vascular flora of the White Mountains of California and Nevada: an updated, synonymized working checklist. Appendix *in* C. A. Hall, Jr. and V. Doyle-Jones (*editors*). The Mary DeDecker Symposium: Plant biology of eastern California. White Mountain Research Station, University of California, Los Angeles.

Morrison, R. B. 1964. Lake Lahontan: Geology of southern Carson Desert, Nevada. U. S. Geological Survey Professional Paper 401. 156 pp.

Morrison, R. B. 1991. Quaternary stratigraphic, hydrologic, and climatic history of the Great Basin, with emphasis on Lakes Lahontan, Bonneville, and Tecopa. Pages 283-320 *in* R. B. Morrison (*editor*). Quaternary nonglacial geology; Conterminous U.S., The Geology of North America, Volume K-2.

Morrow, L. A., and P. W. Stahlman. 1984. The history and distribution of downy brome (*Bromus tectorum*) in North America. *Weed Science* 32, Supplement 1:2-7.

Murphy, D.D. and C.M. Knopp (editors). 2000. Lake Tahoe watershed assessment: Volume I. General Technical Report PSW-GTR-175. USDA Forest Service, Pacific Southwest Research Station, Albany California. 736 pp.

Murphy, D. D., K. E. Freas, and S. B. Weiss. 1990. An environment-metapopulation approach to population viability analysis for a threatened invertebrate. *Conservation Biology* 4:41-51.

National Research Council. 1994. Rangeland health: New methods to classify, inventory, and monitor rangelands. National Academy Press, Washington, D.C. xvi + 180 pp.

The Nature Conservancy. 1994. Spring Mountains National Recreation Area: Biodiversity hotspots and management recommendations. Report to the U.S. Bureau of Land Management, U.S. Fish and Wildlife Service, and U.S.D.A. Forest Service. The Nature Conservancy, Reno, Nevada. 52pp + 23 maps.

Nelson, C. R. 1994. Insects of the Great Basin and Colorado Plateau. Pages 211-238 *in* Natural history of the Colorado plateau and Great Basin. K. T. Harper, L. L. St. Clare, K. H. Thorne, and W. M. Hess (*editors*). University Press of Colorado, Niwot.

Noss, R. F., E. T. LaRoe III, and J. M. Scott. 1995. Endangered ecosystems of the United States: A preliminary assessment of loss and degradation. Biological Report 28, U.S. Department of the Interior, National Biological Service, Washington, D.C. 58 pp.

Nowak, C. L. 1991. Reconstruction of post-glacial vegetation and climate history in western Nevada: Evidence from plant macrofossils in Neotoma middens. M.S. Thesis, Department of Environmental Resources Sciences, University of Nevada, Reno. viii + 69 pp.

Nowak, C. L., R. S. Nowak, R. J. Tausch and P. E. Wigand. 1994. A 30,000 year record of vegetation dynamics at a semi-arid locale in the Great Basin. *Journal of Vegetation Science* 5:579-590.

O'Farrell, T. P. and L. A. Emery. 1976. Ecology of the Nevada Test Site: A narrative summary and annotated bibliography. Report NVO-167. U. S. Department of Energy, National Technical Information Services, U. S. Department of Commerce, Springfield, Virginia.

Ohmart, R. D. 1994. The effects of human-induced changes on the avifauna of western riparian habitats. Pages 273-285 *in* J. R. Jehl, Jr., and N. K. Johnson (*editors*). A century of avifaunal change in western North America. Studies in Avian Biology, Volume 15.

Osborn, G. 1989. Glacial deposits and tephra in the Toiyabe Range, Nevada, U.S.A. *Arctic and Alpine Research* 21:256-267.

Parker, W. S. and E. R. Pianka. 1975. Comparative ecology of populations of the lizard *Uta stansburiana*. *Copeia* 4:615-632.

Patterson, B. D., and W. Atmar. 1986. Nested subsets and the structure of insular mammalian faunas and archipelagos. *Biological Journal of the Linnaean Society* 28:65-82.

Pavlik, B. M. 1985. Sand dune flora of the Great Basin and Mojave deserts of California, Nevada, and Oregon. *Madroño* 32:197-213.

Pavlik, B. M. 1989. Phytogeography of sand dunes in the Great Basin and Mojave deserts. *Journal of Biogeography* 16:227-238.

Pianka, E. R. 1967. On lizard species diversity: North American flatland deserts. *Ecology* 50:1012-1030.

Pickford, G. D. 1932. The influence of continued heavy grazing and of promiscuous burning on spring-fall ranges in Utah. *Ecology* 13:159-171.

Piegat, J. J. 1980. Glacial geology of central Nevada. M.S. Thesis, Department of Geosciences, Purdue University, Indiana.

Platts, W. S. 1990. Managing fisheries and wildlife on rangelands grazed by livestock. Nevada Department of Wildlife. Reno, Nevada.

Porter, S. C., K. L. Pierce, and T. D. Hamilton. 1983. Late Wisconsin mountain glaciation in the western United States. Page 71-111 *in* S. C. Porter (*editor*). Late-quaternary environments of the United States, Volume 1. University of Minnesota Press, Minneapolis.

Pyle, R., M. Bentzien, and P. Opler. 1981. Insect conservation. *Annual Review of Entomology* 26:233-258.

Raven, P. H. 1988. The California flora. Pages 109-137 *in* M. G. Barbour and J. Major (*editors*). Terrestrial vegetation of California. California Native Plant Society, Special Publication Number 9.

Reveal, J. L. 1979. Biogeography of the intermountain region: A speculative appraisal. *Mentzelia* 4:1-92.

Rice, B., and M. Westoby. 1978. Vegetative responses of some Great Basin shrub communities protected against jackrabbits or domestic stock. *Journal of Range Management* 31:28-34.

Robbins, C. S., D. Bystrak, and P. H. Geissler. 1986. The breeding bird survey: Its first 15 years, 1965-1979. U.S. Fish and Wildlife Service, Research Publication No. 157. iii + 196 pp.

Roberts, T. C., Jr. 1990. Cheatgrass: management implications in the 90's. Pages 9-21 *in* E. D. McArthur, E. M. Romney, S. D. Smith, and P.T. Tueller (*editors*). Proceedings: Symposium on cheatgrass invasion, shrub die-off, and other aspects of shrub biology and management, Las Vegas, Nevada, April 5-7, 1989. U.S.D.A. Forest Service, Intermountain Research Station General Technical Report INT-276.

Robinson, S. K. 1988. Reappraisal of the costs and benefits of habitat heterogeneity for nongame wildlife. Pages 145-155 *in* Transactions of the 53rd North American Wildlife and Natural Resources Conference.

Robinson, T. W. 1965. Introduction, spread, and areal extent of saltcedar *(Tamarix)* in the western states. U.S. Geological Survey Professional Paper 491-A. 12pp + 1 plate.

Rogers, G. F. 1982. Then and now: A photographic history of vegetation change in the central Great Basin desert. University of Utah Press, Salt Lake City. 152 pp.

Runeckles, V.C. 1982. Relative death rate: A dynamic parameter describing plant response to stress. *Journal of Applied Ecology* 19:295-303.

Rust, R. W. 1986. New species of *Osmia* (Hymenoptera:Megachilidae) from the southwestern United States. *Entomological News* 97:147-155.

Rust, R. W. 1994. Survey and status of federal category 2 candidate beetle species from Big Dune and Lava Dune in the Amargosa Desert of Nevada. Bureau of Land Management Report for Proposal NV-0546631. 26 pp.

Schmid, R. and M. J. Schmid. 1975. Living links with the past. *Natural History* 84(3):38-45.

Schulz, T. T., and W. C. Leininger. 1991. Nongame wildlife communities in grazed and ungrazed montane riparian sites. *Great Basin Naturalist* 51:286-292.

Scott, J. A. 1986. The butterflies of North America. Stanford University Press, Stanford, California. xii + 583 pp.

Shreve, F. 1942. The desert vegetation of North America. *The Botanical Review* 8:195-246.

Sigler, J. W., and W. F. Sigler. 1994. Fishes of the Great Basin and the Colorado Plateau: past and present forms. Pages 163-208 *in* K. T. Harper, L. L. St. Clair, K. H. Thorne and W. M. Hess, (*editors*). Natural History of the Colorado Plateau and Great Basin. University Press of Colorado, Niwot.

Smith, R. S. U. 1982. Sand dunes in the North American deserts. Page 481-524 *in* G. L. Bender, editor. Reference handbook of the deserts of North America. Greenwood Press, Westport, Connecticut.

Stebbins, G. L. 1974. Flowering plants. Belknap Press, Cambridge, Mass. 399 pp.

Stebbins, R. C. 1974. Off-road vehicles and the fragile desert. *The American Biology Teacher, National Association of Biology Teachers* 36(4,5):203-208; 294-304.

Stebbins, R. C. 1985. A field guide to western reptiles and amphibians. Petersen field guide series. Houghton Mifflin Company. Boston, Massachusetts. xiv + 336 pp.

Stockwell, C. A. 1994. The biology of Walker Lake. Unpublished MS, Biodiversity Research Center, University of Nevada, Reno.

Sudworth, G. B. 1913. Forest atlas. Geographic distribution of North American trees. Part I. Pines. U. S. Department of Agriculture Forest Service. 36 maps (folio).

Sweitzer, R. A. 1990. Winter ecology and predator avoidance in porcupines (*Erethizon dorsatum)* in the Great Basin desert. M.S. Thesis, Department of Biology. University of Nevada, Reno. 64 pp.

Tausch, R. J., and P. T. Tueller. 1990. Foliage biomass and cover relationships between tree- and shrub-dominated communities in pinyon-juniper woodlands. *Great Basin Naturalist* 50:121-134.

Tausch, R. J., N. E. West, and A. A. Nabi. 1981. Tree age and dominance patterns in Great Basin pinyon-juniper woodlands. *Journal of Range Management* 34:259-264.

Terborgh, J. 1989. Where have all the birds gone? Princeton University Press, Princeton, New Jersey. xvi + 207 pp.

Thomas, J. A. 1984. The conservation of butterflies in temperate countries: past efforts and lessons for the future. Pages 333-353 *in* R. I. Vane-Wright and P. R. Ackery (*editors*). The biology of butterflies. Princeton University Press, Princeton, New Jersey.

Thomas, J. A. 1991. Rare species conservation: case studies of European butterflies. Pages 141-197 *in* I. F. Spellerberg, M. G. Morris, and F. B. Goldsmith (*editors*). The scientific management of temperate communities for conservation. 29th Symposium of the British Ecological Society, Blackwell Scientific Publications, Oxford, England.

Thompson, R. S. and J. I. Mead. 1982. Late quaternary environments and biogeography in the Great Basin. *Quaternary Research* 17:39-55.

Tidwell, D. P. 1986. Multi-resource management of pinyon-juniper woodlands: Times have changed, but do we know it? Pages 5-8 *in* R. L. Everett (*editor*), Proceedings: Pinyon-juniper conference. U.S. Forest Service General Technical Report INT-215.

Tueller, P. T., R. J. Tausch, and V. Bostick. 1991. Species and plant community distribution in a Mojave-Great Basin desert transition. *Vegetatio* 92:133-150.

Turner, M. G., W. H. Romme, R. H. Gardner, R. V. O'Neill, and T. K. Kratz. 1993. A revised concept of landscape equilibrium: Disturbance and stability on scaled landscapes. *Landscape Ecology* 8:213-227.

Turner, R. M. 1982. Cold-temperate desertlands. Pages 145-155 *in* D. E. Brown (*editor*), Biotic communities of the American Southwest-United States and Mexico Volume 4: Desert plants.

U.S. Fish and Wildlife Service. 1989. Endangered and threatened wildlife and plants, emergency determination of endangered status for the Mojave population of the desert tortoise. *Federal Register* 54(149):32326.

U.S. Fish and Wildlife Service. 1990. Endangered and threatened wildlife and plants, determination of threatened status for the Mojave population of the desert tortoise. *Federal Register* 55(63):12178-12191.

U.S. Fish and Wildlife Service. 1992. Scoping report: Proposed water acquisition program for Lahontan Valley wetlands under Public Law 101-618. September. i, 24, A1-A3, B1-B4.

U.S. Fish and Wildlife Service. 1994. Endangered and threatened wildlife and plants; determination of critical habitat for the Mojave population of the desert tortoise; Final Rule. *Federal Register* 59:5820-5846.

U.S. Fish and Wildlife Service. 1994. Desert Tortoise (Mojave population) Recovery Plan. U.S. Fish and Wildlife Service, Portland, Oregon. 73 pp. + appendices.

U.S. General Accounting Office. 1993. Livestock grazing on western riparian areas. Gathersburg, Maryland. 44 pp.

Van Devender, T. R. and W. G. Spaulding. 1979. Development of vegetation and climate in the southwestern United States. *Science* 204:701-710.

Vasek, F. C. 1966. The distribution and taxonomy of three western junipers. *Brittonia* 18:350-372.

Vasek, F. C. and M. G. Barbour. 1990. Mojave desert scrub vegetation. Pages 835-867 *in* M. G. Barbour and J. Major (*editors*). Terrestrial vegetation of California. California Native Plant Society, Special Publication Number 9.

Vasek, F. C., H. B. Johnson, and D. H. Elsinger. 1975. Effects of pipeline construction on creosote bush scrub vegetation of the Mojave Desert. *Madroño* 23:1-13.

Vaughan, T. A. 1990. Ecology of living packrats. Pages 14-27 *in* J. L. Betancourt, T. R. Van Devender and P. S. Martin (*editors*). Packrat middens: 40,000 years of biotic change. The University of Arizona Press, Tucson.

Vitousek, P. M. 1986. Biological invasions and ecosystem properties: can species make a difference? Pages 163-176 *in* H. A. Mooney and J. A. Drake (*editors*). Ecology of biological invasions of North America and Hawaii. Springer-Verlag, New York.

Ward, J. V. 1984. Ecological perspectives in the management of aquatic insect habitat. Pages 558-577 *in* V. H. Resh and D. M. Rosenberg (editors). The ecology of aquatic insects. Prager Publishers, Westport, Connecticut.

Weiss, S. B., D. D. Murphy, and R. R. White. 1988. Sun, slope, and butterflies: topographic determinants of habitat quality for *Euphydryas editha*. *Ecology* 69:1486-1496.

Wells, P. V. 1983. Paleobiogeography of montane islands in the Great Basin since the last glaciopluvial. *Ecological Monographs* 53:341-382.

West, N. E. 1988. Intermountain deserts, shrub steppes, and woodlands. Pages 209-230 in M. G. Barbour and W. D. Billings (*editors*). North American terrestrial vegetation. Cambridge University Press, New York.

Wharton, R. A., P. E. Wigand, M. R. Rose, R. L. Reinhardt, D. A. Mouat, H. E. Klieforth, N. L. Ingraham, J. O. Davis, C. A. Fox, and J. T. Ball. 1990. The North American Great Basin: a sensitive indicator of climatic change. Pages 323-359 *in* C. B. Osmond, L. F. Pitelka, and G. M. Hidy (*editors*). Plant biology of the basin and range. Springer-Verlag, New York.

Wheeler, G. C. and J. N. Wheeler. 1986. The ants of Nevada. Natural History Museum of Los Angeles County, Los Angeles, California. vii + 138 pp.

Whisenant, S. J. 1990. Changing fire frequencies on Idaho's Snake River plains: ecological and management implications. Pages 4-10 *in* E. D. McArthur, E. M. Romney, S. D. Smith, and P. T. Tueller (*editors*). Proceedings: Symposium on cheatgrass invasion, shrub die-off, and other aspects of shrub biology and management, Las Vegas, NV, April 5-7, 1989. U. S. Department of

Agriculture Forest Service, Intermountain Research Station General Technical Report INT-276.

Wiederholm, T. 1984. Responses of aquatic insects to environmental pollution. Pages 508-557 *in* V. H. Resh and D. M. Rosenberg, (*editors*). The ecology of aquatic insects. Prager Publishers, Westport, Connecticut.

Wiens, J. A. 1989. The ecology of bird communities. Volume 2. Processes and variations. Cambridge University Press, New York. xii + 316 pp.

Wiens, J. A., and J. T. Rotenberry. 1981. Habitat associations and community structure of shrubsteppe environments. *Ecological Monographs* 51:21-41.

Wigand, P. E., and C. L. Nowak. 1992. Dynamics of northwest Nevada plant communities during the last 30,000 years. Pages 40-61 *in* C. A. Hall, Jr., V. Doyle-Jones, and B. Widawski (*editors*). The history of water: Eastern Sierra Nevada, Owens Valley, White-Inyo Mountains. White Mountain Research Station Symposium Volume 4.

Wigand, P. E., M. L. Hemphill, and S. M. Patra. 1994. Late Holocene climate derived from vegetation history and plant cellulose stable isotope records from the Great Basin of western North America. Pages 2574-2583, *in* Proceedings of the High-Level Radioactive Waste Management Conference & Exposition, May 22-26, 1994. Las Vegas, Nevada.

Wilcox, B. A., D. D. Murphy, P. R. Ehrlich, and G. T. Austin. 1986. Insular biogeography of the montane butterfly faunas in the Great Basin: Comparison with birds and mammals. *Oecologia* 69:188-194.

Williams, T. R. 1982. Late Pleistocene lake level maxima and shoreline deformation in the basin and range province, western United States. M.S. Thesis, Earth Sciences Department. Colorado State University, Fort Collins, Colorado. 52 pp.

Wright, H. E., Jr. 1983. Introduction. Pages xi-xvii *in* H. E. Wright, Jr. (*editor*). Late-quaternary environments of the United States, Volume 2: The Holocene. University of Minnesota Press, Minneapolis.

Yensen, D. I. 1981. The 1900 invasion of alien plants into southern Idaho. *Great Basin Naturalist* 41:176-183.

Young, J. A., and R. A. Evans. 1978. Population dynamics after wildfires in sagebrush grasslands. *Journal of Range Management* 31:283-289.

Young, J. A., R. A. Evans, and J. Major. 1972. Alien plants in the Great Basin. *Journal of Range Management* 25(3)194-201.

Young, J. A., R. A. Evans, B. A. Roundy, and J. A. Brown. 1986. Dynamic landforms and plant communities in a pluvial lake basin. *Great Basin Naturalist* 46:1-21.

Zeveloff, S. I. 1988. Mammals of the intermountain west. University of Utah Press, Salt Lake City. xxiv + 365 pp.

Student Study Aids

Terms Used in Nevada Environmental Issues

David Charlet

Word Games for the Precocious Student

To pass a college course, you must master its vocabulary. In all disciplines, from auto mechanics to calculus, from architecture to environmental science, there is a set of words with special meanings, or vocabulary, that workers in that discipline use to communicate precisely and concisely with each other. Sometimes, a common word or pair of words has a special meaning in a discipline. For instance, *community* means one thing to most people and something else to an ecologist. Both definitions are correct, but in this class, you must know what an ecologist means by community. When you take a general studies college course, you investigate that discipline to gain an understanding of its basic concepts, and these concepts are embedded in its language. To understand the lectures, your readings, to write a good term paper, and even to read the exam questions (in short, to pass this course), you need to know the definitions of all these terms; you must be comfortable with the vocabulary.

From the first day of class through the end, you should work on your vocabulary words. The following section contains vocabulary worksheets and a place to put all of your vocabulary information in one place. What follows is a list of vocabulary words and a list of acronyms and other abbreviations used in the text (***Vocabulary List*** and ***Acronym and Abbreviation List***). These lists repeat twice, so that both the vocabulary and the acronyms and abbreviations have a Worksheet and a Self-Test section. Following each word in the Worksheets are boxes where you can write definitions and root word meanings for each vocabulary word and acronym in the book. Here are three exercises you can do using your list, this textbook, your primary textbook, and your good college dictionary.

Exercise One: Definitions and Index

As soon as you encounter a word that you do not understand, look to see if it is in the ***Vocabulary List*** (the three-column list of words immediately following).

If the word is not in the list, then it may not be a technical term, just a word you didn't know before. Regardless, look up the word and write the word and its definition down in the ***Vocabulary Index and Worksheet*** after the three-column list of words. Also, in the box provided to the left of the word, write down the page number where you found the word. In this way, you are building your own index for the book.

If the word is on the ***Vocabulary List***, then look to see if the word is in a glossary at the end of the chapter. If so, write down the definition from the glossary in the space following the word, and write down the page on which you found it. Then, look up the word in your dictionary and write down the appropriate definition. Perhaps your instructor defined the term in class. Write down all the definitions you find for the term in the space provided by the word. Are there differences in the definitions? Are there words you do not understand in the definitions? If so, you should look up these definitions too, and write them down as new words and definitions.

Exercise Two: Root Meanings

Most dictionaries provide meanings for the root words, prefixes, and suffixes in scientific words. In the ***Vocabulary Index and Worksheet***, in the box to the right of the word, divide the word up into its component parts and write the meanings out for these parts. A limited set of root words, suffixes and prefixes are found in scientific words, so by learning the meanings of the parts, you can gain mastery over the vocabulary more quickly than if you try to memorize whole words. Also, when you encounter new words, you can often at least get a general idea of what the word means by knowing the meaning of its parts.

For instance, the first word on your list is **abiotic**. This word is made of 3 parts, **a – bio – tic**. The "a" means without whatever follows it. The "tic" at the end means pertaining to whatever preceded it. "Bio" means life. And so, a-bio-tic means, "without life." You will find the prefix "a" frequently in scientific words — if it is in front of a term that you know, but you have never seen it with the "a" in front of it, do not fear — you know it means "without" whatever that term is that you know! Sometimes the term we modify with an "a" as a prefix begins with a vowel, like **aerobic** (*aero-* = air; *-bic* = pertaining to). We don't put the "a" in front of another vowel, so we add use "an" instead of "a." So if we mean "without air" we write **anaerobic.** (Don't ask me why the –ic ending has a "t" before it in abiotic and a "b" before it in aerobic, because I don't know. The key point here is that "-ic" means pertaining to, and since aero and bio both end in vowels and "–ic" begins with a vowel, these are separated by a consonant.) Try breaking up your words in this way, and learning these word parts. You may be amazed at how many you find in words you use every day!

You can follow the steps above in filling out the definitions for the acronyms and abbreviations in the ***Acronym and Abbreviation Index and Worksheet***. If you do not understand one of the words that one of the letters in the acronym stands for, then look to see if it is on the vocabulary list. If not, then you should treat it as if it is another vocabulary word, and add it to your ***Vocabulary Index and Worksheet***.

Exercise Three: Self-Testing

A good way to test your own understanding before the final exam is to use the ***Vocabulary Self-Test*** (the third and last time the terms are listed, and the second time the terms are placed near boxes). Use the boxes provided to write out definitions for these words in pencil without looking them up. While working on the list, if you find words you do not understand, then you should look them up again. However, do not write them down until you can do so correctly from memory. Try to get through the whole list in this way. Once you get through the whole list filling definitions where you can, grade yourself. That is, look up each word again and compare the definition given by the reference to what you wrote down from memory. Did you remember the essential points of the definition? If your definition was correct, leave it, but if not, then erase your answer.

Once you have gone through all of your answers, then you are ready to start again. Begin at the top, and go to each of the blank definition boxes by the words and fill in the definitions as you can. Then repeat the grading process. Go through the list in this way until you have filled in the correct definitions for all the terms. Use this same process to test your own mastery of the acronyms and abbreviations used in the text in the ***Acronym and Abbreviation Self-Test***. Once you have completed all three exercises, you will have much more confidence in the course material.

Vocabulary List

abiotic
acronym
affluence
alfalfa
alpine
ammonium perchlorate
amphibian
annual
anthrocentric
anthropogenic
arsenic
atmosphere
australis
bacteria
batholith
bioaccumulate
biocentric
biodiversity
biological soil crust
biology
biomass
biome
biotic
bird
birth rate
boreal
broadleaf
bronchitis
carbon cycle
carbon dioxide
carbon monoxide
carcinogen
carrying capacity
case-control
Cenozoic
chemical
chemical agents
chemical weathering
chemistry
chronic
clastic
climate
coal
co-generation
community
cone
conifer
conservation of energy
country rock
critical thinking
cryogenic
deciduous
demographic transition
denitrification
desert
desertification
developed nations
developing nations
dominant plant
drought deciduous
ecocentric
ecology
economy
ecosystem
ecotone
effluent
element
embrittlement
emigration
emphysema
endemic
endocrine-disrupting
compound
energy
environment
environmental
epidemiology
environmental resistance
epidemiology
equilibrium
estrogen
evergreen
exoskeleton
exponential

extinction
extirpated
extrusive rock
fallacy
fauna
feral
fish
flora
flower
foliation
forest
fossil
freshwater ecosystem
geography
geology
glacier
glucose
gneiss
graminoid
granite
grassland
Great Basin
habitat
hemoglobin
herbs
hydrocarbon
hydroelectric
hydrologic cycle
hydroponics
hydrothermal
idle
igneous
infectious agents
interglacial
intrusive rock
immigration
immunotoxin
invertebrate
kilowatt hour
kinetic energy
landscape
lethal dose
lethality
leukemia
life form
lifestyle
limestone
lithified
magma
mainstream smoke
mammal
meadow
mechanical weathering
megafauna
Mesozoic
metabolism
metamorphic
micron
microphyllous
migration
mineral
Mojave Desert
molecule
morbity
mortality
needleleaf
neurotoxin
nitrogen fixation
nitrogen oxides
non-attainment
nontoxic
oil
oxygen
ozone
Paleozoic
particulate
perennial
petrified
phenology
photosynthesis
photovoltaic cell
physical agents
physics
physiognomy
Pleistocene
Pliocene
pluton
polar
pollen
pollution
polygon
population
population density
Precambrian
primary pollutant
recycle

reflectance
reproduction
reptile
respiration
rhyolite
sandstone
seasonally deciduous
secondary pollutant
sedimentary
shrub
shrubland
sidestream smoke
siliceous
specificity
stomate
stomatal conductance
strain
strass
subalpine
superpositioning
surface albedo
technology
tectonic
temperate
teratogen
thermodynamics
total fertility rate
toxicology
toxin
transpiration
tree
tropical
tuff
tundra
ubiquitous
uniformitarianism
vaccination
vegetation
vegetation structure
vertebrate
virus
volatile organic compound
volcanic
weather
weathering
wetland
woodland
work
xerophytic

Acronym and Abbreviation List

AIDS

ALL

AML

BIA

BLM

BP

BRRC

bu.

CDC

CEO

CO

CO_2

cwt.

DIY

DNA

DOD

DOE

EDC

EIR

ELF

EPA

ESA

ETS

FACE

GAP

GIS

HIV

ICRISAT

LD_{50}

MCL

MID

MSDS

MSHCP

mya

NAAQS

NASS

NCR

NDEP

NDFF

NDOW

NIMBY

NFT

NO_x

NPS

NWR

O_2

O_3

OPEC

PM

ppb

ppm

SIDS

TID

TSH

USDA

USDA FS

USFWS

VOC

Vocabulary Index and Worksheet

pg	Word	Root words	Definition
	abiotic		
	acronym		
	affluence		
	alfalfa		
	alpine		
	ammonium perchlorate		
	amphibian		
	annual		
	anthrocentric		

	anthropogenic		
	arsenic		
	atmosphere		
	australis		
	bacteria		
	batholith		
	bioaccumulate		
	biocentric		
	biodiversity		
	biological soil crust		
	biology		

	biomass		
	biome		
	biotic		
	bird		
	birth rate		
	boreal		
	broadleaf		
	bronchitis		
	carbon cycle		
	carbon dioxide		
	carbon monoxide		

	carcinogen		
	carrying capacity		
	case-control		
	Cenozoic		
	chemical		
	chemical agents		
	chemical weathering		
	chemistry		
	chronic		
	clastic		
	climate		

	coal		
	co-generation		
	community		
	cone		
	conifer		
	conservation of energy		
	country rock		
	critical thinking		
	cryogenic		
	deciduous		
	demographic transition		

	denitrification		
	desert		
	desertification		
	developed nations		
	developing nations		
	dominant plant		
	drought deciduous		
	ecocentric		
	ecology		
	economy		
	ecosystem		

	ecotone		
	effluent		
	element		
	embrittlement		
	emigration		
	emphysema		
	endemic		
	endocrine-disrupting compound		
	energy		
	environment		
	environmental epidemiology		

	environmental resistance		
	epidemiology		
	equilibrium		
	estrogen		
	evergreen		
	exoskeleton		
	exponential		
	extinction		
	extirpated		
	extrusive rock		
	fallacy		

	fauna		
	feral		
	fish		
	flora		
	flower		
	foliation		
	forest		
	fossil		
	freshwater ecosystem		
	geography		
	geology		

	glacier		
	glucose		
	gneiss		
	graminoid		
	granite		
	grassland		
	Great Basin		
	habitat		
	hemoglobin		
	herbs		
	hydrocarbon		

	hydroelectric		
	hydrologic cycle		
	hydroponics		
	hydrothermal		
	idle		
	igneous		
	infectious agents		
	interglacial		
	intrusive rock		
	immigration		
	immunotoxin		

	invertebrate		
	kilowatt hour		
	kinetic energy		
	landscape		
	lethal dose		
	lethality		
	leukemia		
	life form		
	lifestyle		
	limestone		
	lithified		

	magma		
	mainstream smoke		
	mammal		
	meadow		
	mechanical weathering		
	megafauna		
	Mesozoic		
	metabolism		
	metamorphic		
	micron		
	microphyllous		

	migration		
	mineral		
	Mojave Desert		
	molecule		
	morbity		
	mortality		
	needleleaf		
	neurotoxin		
	nitrogen fixation		
	nitrogen oxides		
	non-attainment		

	nontoxic		
	oil		
	oxygen		
	ozone		
	Paleozoic		
	particulate		
	perennial		
	petrified		
	phenology		
	photosynthesis		
	photovoltaic cell		

	physical agents		
	physics		
	physiognomy		
	Pleistocene		
	Pliocene		
	pluton		
	polar		
	pollen		
	pollution		
	polygon		
	population		

	population density		
	Precambrian		
	primary pollutant		
	recycle		
	reflectance		
	reproduction		
	reptile		
	respiration		
	rhyolite		
	sandstone		
	seasonally deciduous		

	secondary pollutant		
	sedimentary		
	shrub		
	shrubland		
	sidestream smoke		
	siliceous		
	specificity		
	stomate		
	stomatal conductance		
	strain		
	strass		

	subalpine		
	superpositioning		
	surface albedo		
	technology		
	tectonic		
	temperate		
	teratogen		
	thermodynamics		
	total fertility rate		
	toxicology		
	toxin		

	transpiration		
	tree		
	tropical		
	tuff		
	tundra		
	ubiquitous		
	uniformitarianism		
	vaccination		
	vegetation		
	vegetation structure		
	vertebrate		

	virus		
	volatile organic compound		
	volcanic		
	weather		
	weathering		
	wetland		
	woodland		
	work		
	xerophytic		

Acronym and Abbreviation Index and Worksheet

pg	Acronym or abbreviation	Definition
	AIDS	
	ALL	
	AML	
	BIA	
	BLM	
	BP	
	BRRC	
	CDC	
	CEO	

	CO	
	CO_2	
	DIY	
	DNA	
	DOD	
	DOE	
	EDC	
	EIR	
	ELF	
	EPA	
	ESA	

	ETS	
	FACE	
	GAP	
	GIS	
	HIV	
	ICRISAT	
	LD_{50}	
	MCL	
	MID	
	MSDS	
	MSHCP	

	mya	
	NAAQS	
	NASS	
	NCR	
	NDEP	
	NDFF	
	NDOW	
	NIMBY	
	NFT	
	NO_x	
	NPS	

	NWR	
	O_2	
	O_3	
	OPEC	
	PM	
	ppb	
	ppm	
	SIDS	
	TID	
	TSH	
	USDA	

	USDA FS	
	USFWS	
	VOC	

Vocabulary Self-Test

Word	Definition
abiotic	
acronym	
affluence	
alfalfa	
alpine	
ammonium perchlorate	
amphibian	
annual	
anthrocentric	

anthropogenic	
arsenic	
atmosphere	
australis	
bacteria	
batholith	
bioaccumulate	
biocentric	
biodiversity	
biological soil crust	
biology	

biomass	
biome	
biotic	
bird	
birth rate	
boreal	
broadleaf	
bronchitis	
carbon cycle	
carbon dioxide	
carbon monoxide	

carcinogen	
carrying capacity	
case-control	
Cenozoic	
chemical	
chemical agents	
chemical weathering	
chemistry	
chronic	
clastic	
climate	

coal	
co-generation	
community	
cone	
conifer	
conservation of energy	
country rock	
critical thinking	
cryogenic	
deciduous	
demographic transition	

denitrification	
desert	
desertification	
developed nations	
developing nations	
dominant plant	
drought deciduous	
ecocentric	
ecology	
economy	
ecosystem	

ecotone	
effluent	
element	
embrittlement	
emigration	
emphysema	
endemic	
endocrine-disrupting compound	
energy	
environment	
environmental epidemiology	

environmental resistance	
epidemiology	
equilibrium	
estrogen	
evergreen	
exoskeleton	
exponential	
extinction	
extirpated	
extrusive rock	
fallacy	

fauna	
feral	
fish	
flora	
flower	
foliation	
forest	
fossil	
freshwater ecosystem	
geography	
geology	

glacier	
glucose	
gneiss	
graminoid	
granite	
grassland	
Great Basin	
habitat	
hemoglobin	
herbs	
hydrocarbon	

hydroelectric	
hydrologic cycle	
hydroponics	
hydrothermal	
idle	
igneous	
infectious agents	
interglacial	
intrusive rock	
immigration	
immunotoxin	

invertebrate	
kilowatt hour	
kinetic energy	
landscape	
lethal dose	
lethality	
leukemia	
life form	
lifestyle	
limestone	
lithified	

magma	
mainstream smoke	
mammal	
meadow	
mechanical weathering	
megafauna	
Mesozoic	
metabolism	
metamorphic	
micron	
microphyllous	

migration	
mineral	
Mojave Desert	
molecule	
morbity	
mortality	
needleleaf	
neurotoxin	
nitrogen fixation	
nitrogen oxides	
non-attainment	

nontoxic	
oil	
oxygen	
ozone	
Paleozoic	
particulate	
perennial	
petrified	
phenology	
photosynthesis	
photovoltaic cell	

physical agents	
physics	
physiognomy	
Pleistocene	
Pliocene	
pluton	
polar	
pollen	
pollution	
polygon	
population	

population density	
Precambrian	
primary pollutant	
recycle	
reflectance	
reproduction	
reptile	
respiration	
rhyolite	
sandstone	
seasonally deciduous	

secondary pollutant	
sedimentary	
shrub	
shrubland	
sidestream smoke	
siliceous	
specificity	
stomate	
stomatal conductance	

subalpine	
superpositioning	
surface albedo	
technology	
tectonic	
temperate	
teratogen	
thermodynamics	
total fertility rate	
toxicology	
toxin	

transpiration	
tree	
tropical	
tuff	
tundra	
ubiquitous	
uniformitarianism	
vaccination	
vegetation	
vegetation structure	
vertebrate	

virus	
volatile organic compound	
volcanic	
weather	
weathering	
wetland	
woodland	
work	
xerophytic	

Acronym and Abbreviation Self-Test

Acronym or abbreviation	Definition
AIDS	
ALL	
AML	
BIA	
BLM	
BP	
BRRC	
CDC	
CEO	
CO	
CO_2	
DIY	
DNA	
DOD	

DOE	
EDC	
EIR	
ELF	
EPA	
ESA	
ETS	
FACE	
GAP	
GIS	
HIV	
ICRISAT	
LD_{50}	
MCL	
MID	
MSDS	
MSHCP	
mya	

NAAQS	
NASS	
NCR	
NDEP	
NDFF	
NDOW	
NIMBY	
NFT	
NO_x	
NPS	
NWR	
O_2	
O_3	
OPEC	
PM	
ppb	
ppm	
SIDS	

TID
TSH
USDA
USDA FS
USFWS
VOC